MEIOSIS IN THE POLLEN MOTHER CELLS
OF SOME CANNAS

MEIOSIS IN THE POLLEN MOTHER CELLS OF SOME CANNAS

PROEFSCHRIFT TER VERKRIJGING VAN DEN GRAAD VAN DOCTOR IN DE WIS- EN NATUURKUNDE AAN DE RIJKS UNIVERSITEIT TE UTRECHT OP GEZAG VAN DEN RECTOR MAGNIFICUS DR. C. W. VOLLGRAFF, HOOGLEERAAR IN DE FACULTEIT DER LETTEREN EN WIJSBEGEERTE, VOLGENS BESLUIT VAN DEN SENAAT DER UNIVERSITEIT TEGEN DE BEDENKINGEN VAN DE FACULTEIT DER WIS- EN NATUURKUNDE TE VERDEDIGEN OP MAANDAG 9 DECEMBER 1935, DES NAMIDDAGS TE 3 UUR

DOOR

FREDERIK JAN MAURITS OFFERIJNS
GEBOREN TE DEVENTER

'S-GRAVENHAGE
MARTINUS NIJHOFF
1935

ISBN 978-94-015-2061-4 ISBN 978-94-015-3247-1 (eBook)
DOI 10.1007/978-94-015-3247-1

AAN MIJN VROUW

Bij het voltooien van dit proefschrift wil ik gaarne een blik terug werpen:

Het zij mij vergund, hier in eerbiedige herinnering vóór alles te getuigen van de dankbaarheid, die ik gevoel jegens wijlen Prof. Dr. F. A. F. C. WENT, den leermeester, wiens belangstelling in mijn werk en leven mij vele malen tot steun is geweest.

Met groote waardeering en erkentelijkheid gedenk ik het ouderlijk huis, waar aanleg en omgeving samenwerkten, om liefde voor de natuur te ontwikkelen en haar schoonheid te leeren genieten.

Met bewondering en eerbied denk ik terug aan het onderwijs van wijlen Dr. J. J. LE ROY, die niet slechts op degelijke en boeiende wijze de „levensverschijnselen van den mensch" behandelde, maar wiens wijsgeerige geest daarbij belangstelling wist te wekken voor den „ontwikkelingsgang van het menschelijk denken".

Zeergeleerde HOOGENRAAD, ongetwijfeld hebt Gij indertijd den grootsten invloed op mij uitgeoefend. Niet slechts door Uw bekwaamheden als docent, maar evenzeer door Uw gaven als mensch, zijt Gij in staat de lust tot eigen waarneming en onderzoek te wekken en te leiden bij Uw leerlingen, zonder dat zij daarbij eigen aard behoeven te verloochenen. Voor Uw voortdurende belangstelling en vriendschap, die ik reeds in zoo ruime mate genoten heb, zal ik steeds dankbaar blijven.

Hooggeleerde JORDAN, KONINGSBERGER, NIERSTRASZ, PULLE en RUTTEN, zeer erkentelijk ben ik voor wat ik van U mocht leeren of voor de welwillendheid, die Gij mij wel hebt betoond. Hooggeleerde PULLE, ten zeerste waardeer ik het, dat Gij steeds bereid zijt geweest mij met Uw raad bij te staan en dat Gij mij bij het bewerken van dit proefschrift in enkele moeilijkheden van advies hebt willen dienen.

Aan Uw practica in de Protistologie, zeergeleerde ENTZ, bewaar ik nog de meest aangename herinneringen.

Hooggeleerde HONING, hooggeachte Promotor, voor zeer veel ben ik U dank verschuldigd. Dat Gij mij op zoo onbekrompen wijze materiaal en hulpmiddelen verschaft hebt, om dit onderzoek te kunnen uitvoeren, waardeer ik meer dan ik zeggen kan; dat Gij, ondanks Uw drukke bezigheden, steeds tijd hebt gevonden voor kritische besprekingen, die aan het werk ten goede zijn gekomen, stel ik op hoogen prijs. De uren in Uw gastvrij laboratorium doorgebracht, zullen voor mij dan ook van groote beteekenis blijven!

Waarde Berrington, wees verzekerd, dat Uw hulp bij de correctie van de Engelsche tekst ten zeerste door mij wordt gewaardeerd; ook voor Uw medewerking, waarde vriend Reilingh, ben ik nog zeer erkentelijk.

Ook U, zeergeleerde Gerrits, ben ik erkentelijk voor de welwillendheid, mij betoond bij de regeling der lessen en voor Uw belangstelling in dit werk.

Zeergeleerde Mej. Polak, gaarne richt ik ook tot U een woord van dank voor de vriendelijke wijze, waarop Gij mij steeds de literatuur hebt willen verschaffen, die ik noodig had.

Waarde Jansen, Knoop en van de Peppel; Gij hebt mij vele malen door Uw hulp en voorkomendheid het werken in tuin en laboratorium vergemakkelijkt; daarvoor zeg ik U hartelijk dank.

Had ik niet de voortdurende vaardige hulp gehad van mijn vrouw bij de technische bewerking van het materiaal, dan was het nu verkregen resultaat nog niet bereikt; haar blijvende belangstelling in den voortgang van het onderzoek heeft mij de uren van gemeenschappelijken arbeid onvergetelijk gemaakt.

MEIOSIS IN THE POLLEN MOTHER CELLS OF SOME CANNAS

by

F. J. M. OFFERIJNS
(*Received for publication Oct. 25th 1935*)

TABLE OF CONTENTS

I. INTRODUCTION

The researches have been performed on some representatives of the genus *Canna*.

The intention of the author was:

a. to describe the general course of the meiosis of the pollen mother cells,

b. to compare this course of different forms which have been studied,

c. to trace how far with the representatives of the genus *Canna* particularities appear in the meiotic phenomena which are of importance for the contemplation of the method of pairing of the chromosomes.

Through former cytological researches by HONING (1928) the expectation was roused that in this material something might be found for a critical inspection of the problem about para- and telosyndesis.

The cytological literature with regard to species and varieties of the genus *Canna* is not very extensive, most documents about that matter are of a systematical and morphological nature, especially, in connection with the remarkable structure of the flower (COSTERUS, 1916, 1920). After 1900, also this genus was used for genetical researches; for several forms the number of chromosomes was examined, but so far attempts to make a connection between cytological and genetical occurrences have not been made (apart from a single exception, BELLING, 1921, 1925, HONING, 1915, 1923, 1928).

The material used for the research written down here was obtained from the Canna-cultures of Prof. HONING, who for his genetical researches grows a number of forms in the greenhouses of the „LABORATORY FOR GENETICS OF THE UNIVERSITY COLLEGE OF AGRICULTURE", WAGENINGEN. I would tender him my best thanks for his kind permission to collect material in his experimental garden.

The original plants grown there came from various parts of the world; the cytologically worked material was collected from genetically well-known plants.

II. HISTORICAL REVIEW

KRÄNZLIN gives, in ENGLER, Das Pflanzenreich (Band IV, 47,

Cannaceae I, 1912) a historical review of the research of the genus *Canna*, which is worth studying for everybody who is occupied with these plants.

Very likely Cannas were already grown in Europe at the end of the 16th century. The first clear statement about that occurs in BAUHIN's Pinax Theatri Botanici which was printed in 1623. These plants are also mentioned by TOURNEFORT, PISO, RHEEDE and RUMPHIUS.

For the first time three species are mentioned in LINNÉ's „Species Plantarum" viz. *Canna angustifolia*, *Canna glauca* and *Canna indica*; the latter with a diagnosis of little significance and with a collection of varieties representing almost as many species, so that KRÄNZLIN remarks: „Vom allerersten Anfang an ein Stein des Anstosses".

In the 19th century the interest for *Canna* is evidently exceptionally great, as well from the side of scientific researchers as from the side of the practical growers.

Many forms were collected and described in the middle of last century, of which the herbar-materials are deposited in Berlin—Dahlem at present. These collections are worked by KRÄNZLIN and reduced by him to 51 species: 43 species from America, 6 from Asia, 1 from Africa and 1 from unknown origin.

Suppose we take this number of species as a basis and state that under the influence of the interest of the horticultural artists, hybrids have been grown and again hybrids of hybrids, then it is possible to understand that from cultivated forms which at present are scattered all over the tropical and subtropical parts of the world as „garden-plants", the exact systematical place cannot always be directly and easily determined. For it must be added that many growers controlled insufficiently or not at all what had resulted from the crossing; the statements of the names of crossed species are often altogether wrong (sometimes even purposely(?) misleading, according to COSTERUS, 1916*a*). It is quite well possible that so-called „wild" forms are escaped garden-plants and of hybrid origin. Undoubtedly forms of a very different origin, bearing the same name, have been in cultivation, and so it may be explained that different statements are to be found for the „same" kind of plants, with regard to the number of chromosomes. As an example: for *Canna indica* are to be found for the haploid number the statements 3, 8, 9; for the diploid number 9+9/2, 27/2 (GAISER, 1926, 1930*a*, 1930*b*; TISCHLER,

1927). They were either different types or different „indica"-hybrids.

Some authors mention *Canna* sp., *Canna* hybrids, „garden-varieties", without further statements, but that is why it is not always possible to compare precisely their results.

III. CLASSIFICATION

As already pointed out the present classification of the species in the genus *Canna* is due to the research of KRÄNZLIN (1912) and later writers have adopted his diagnoses and tables (WINKLER, 1930; WETTSTEIN, 1935).

In order to make comparison of examined plants possible, it is now desirable to use KRÄNZLIN's tables as a basis for the description and the denomination of the species.

The forms, cytologically described (in this investigation) belong to the subgenus *Eucanna*, which is characterised by the presence of two or three petal-like staminodes in addition to the labellum. This subgenus is divided into:

Sectio I, BIALATAE (with 2 staminodes) and

Sectio II, TRIALATAE (with 3 staminodes).

Canna humilis BOUCHÉ (KRÄNZLIN, 1912, p. 43) received from the botanical gardens at Montevideo, is a species with two staminodes and with totally green and glabrous leaves. The plants used slightly differ from the description given by KRÄNZLIN, but yet in my opinion this form must be ranked among the species. The variability of the characteristics of cultivated forms is apparently very considerable, especially as regards the quantitative properties (cf. HONING, 1923, p. 4). KRÄNZLIN (l.c. p. 12) says:

„Die Variabilität der Färbung bei Exemplaren derselben Art in Regeln zu fassen, muss ich für ein zunächst wenigstens aussichtsloses Unterfangen halten". The leaf-blades of *Canna humilis* are light-green, white-margined, ovate. (70 cm. × 23 cm.) The bracts are light-green with pinkish margin. (±30 mm. long, ±15 mm. br.) Ovary: green with red. Sepals: crimson to brownish-red, the middle greenish, the outer surface glaucous. Petals: carmine, upwards light-red. (40 mm. long.) The inner surface yellow-green, striped. Labellum: the underside of the curved part orange with red margin, the upper side is yellow-orange and red-spotted, the apex is red. Staminodes: two, bilobate, nearly 6 cm. long, narrow, brilliant vermilion-red (fig. 1 and fig. 2).

Canna discolor LINDL. (KRÄNZLIN, 1912, p. 38). The variety examined is also a form, which does not altogether correspond with the compared descriptions. KRÄNZLIN attaches great value to the number of petal-like staminodes, for he uses it as a basis for the classification of the subgenus *Eucanna*; according to this classification the mentioned variety must be considered to belong to the *Bialatae*, although very seldom (twice out of some thousands of flowers) a third staminode has been observed (by HONING) noticeably but not very much smaller than the outer ones. The exceptionally great development of the dark-red colour in stems, leaf-sheaths, leaf-margins, bracts, inflorescences, flowers and ovaries, is the most prominent of all the characteristics of this variety. Leaves: ovate, totally glabrous. Leaf blades: dark-green, shiny, with dark-red margin and middle-vein, the top very acute. All parts of the inflorescences are totally dark-red. Bracts: glaucous (25 mm.l., 20 mm. br.) Sepals: dark-red, (15—20 mm. l., narrow). Petals: linear lanceolate, the outher surface deep-red, the inner side slightly greenish, (35—40 mm. l.) Staminodes: generally 2, bilobate, deep-red. Labellum: red, the middle flesh-coloured, red-spotted (fig. 1 and fig. 2). It is, of course, quite well possible that the characteristics of this variety, which differ from those of the pure *Canna discolor*, are due to a previous crossing with a trialate species (such as *Canna indica* L).

Canna lutea MILL (KRÄNZLIN, 1912, p. 39) also received from the botanical gardens at Montevideo, is a bialate species with light green leaves, the leaf-sheaths and leaf-blades are glabrous (the form is that of *Canna discolor*). Bracts: light-green, (30 mm. l.), glaucous. Ovary: light-green. Sepals: light-green (14 mm. l.), glaucous. Petals: light-yellow, towards the top greenish, (45 mm. l.). Labellum: light-yellow, upper side red-spotted. Staminodes: 2, the same colour as the labellum, both sides with light-red spots and stripes; top bilobate or sometimes nearly entire (The appearance of the flower may be judged from the photograph, fig. 1). The above mentioned Cannas (*C. humilis, discolor* and *lutea*) are, no doubt, closely related.

Canna glauca L. (with the varieties „JAVA", „BOLIVIA", „MONTEVIDEO" and „PURE YELLOW"), a species of the Trialatae.

For the description of these forms we may refer to the publications by HONING and to the photograph, fig. 1, showing the form of the flowers of the varieties examined. It was not possible to identify these forms more accurately with forms given by KRÄNZLIN; therefore the names mentioned by HONING, are used.

FIG. 1. Flowers of *Canna discolor, humilis, lutea* and *Canna glauca*: Java, Bolivia, Montevideo and Pure yellow.

IV. THE EARLIER CYTOLOGICAL INVESTIGATIONS

A historical review of the literature about the cytology of the reproduction-organs can be relatively short here, because in the preceding century only little or nothing at all was written with regard to the research of the chromosomes in species of the genus *Canna* (cf. SCHÜRHOFF, 1926; SCHNARF, 1931).

Before 1900 the cytological publications chiefly relate to the embryogenesis in general, but they do not yet contain information about the origin and the division of the chromosomes.

The first essay in which something is mentioned about the number of chromosomes and the reduction-division of *Canna indica*, appeared in 1900. It is a publication by KARL M. WIEGAND: „The development of the embryosac in some monocotyledonous plants". He describes and compares the development of the embryosac in *Convallaria majalis* L., *Potamogeton foliosus* RAF. and *Canna indica* L. His conclusion is: „The development in *Canna* was found to be nearly normal". In the summary of his paper he says with regard to *Canna*:

„The number of chromosomes in the vegetative divisions is six. When passing to the poles at the heterotypic division there were still six; but later the second division showed only three as the reduced number. Probably the segmentations for both divisions occur during the prophase of the heterotypic division. This number is one of the smallest yet found in vegetable tissue".

The illustrations belonging to this paper are clear and do not suggest any doubt about the number of chromosomes. An anaphase-representation clearly shows the separation of two sets of six chromosomes each. About the development of the chromosomes WIEGAND did not see much, which appears from his words: „The nuclei and chromosomes in *Canna* are so small that little could be done toward working out the segmentation of the latter". Possibly he means this as compared with the chromosomes of *Convallaria* and *Potamogeton*; it is not to be stated with certainty because neither from the illustrations themselves does it appear nor is in the text the enlargement of the figures mentioned. Though the prophase stages presented difficulties in observation, this was not the case with the following ones, for he says further on: „the ordinary process of division assures six daughter chromosomes for each resulting nucleus. This count was made many times with great ease, owing to the small number of segments and always with the same result".

Still his statement must be assumed with some reservation for he also says that he found more than six chromosomes in the archesporial nucleus. Moreover it must be taken into consideration that the number of countings of the material was not large: „Only two spindles were found representing the heterotypic division and these both showed the globular daughter segments on their way to the poles (anaphase). The four counts here made gave in every case the number of six, instead of three as one would expect after reduction". And he only saw one or two sections with the second division.

It is not to be settled whether WIEGAND dealt with a deviating form of *Canna indica* or with abnormal material; anyhow, the observations mentioned by him were not affirmed later on. Soon after the appearance of the discussed essay a doubt was expressed about the truth of the statements made by WIEGAND.

In 1903 M. KOERNICKE gave a critical report about the state of the cytology, the knowledge of protoplasma and nucleus: „Der heutige Stand der pflanzlichen Zellforschung". KOERNICKE in order to control the result of WIEGAND, made the same investigation once more. „Die eigentümliche Art und Weise, auf welche sich der Reduktionsvorgang in den Embryosackmutterzellen von *Canna* nach WIEGAND vollziehen soll, veranlassten mich zu einer Prüfung der Angaben". „Die Resultate meiner Nachprüfung, welche ich hier anfügen möchte waren folgende: Zunächst traten mir bei Beginn der ersten Teilung acht längsgespaltene Chromosomen entgegen, die sich stark verkürzten. Im Äquator der ersten Spindel finden wir die acht Doppelklümpchen wieder, deren Hälften nach den Polen auseinanderweichen. Bei dem zweiten Teilungsschritt fand sich die Achtzahl der Chromosomen wieder. Dieselbe Zahl trat mir auch in den beiden entsprechenden Teilungsschritten der Pollenmutterzellen, die ich daraufhin studierte entgegen. Nach den Beobachtungen die ich bei *Canna* machte liegt für mich die Annahme nahe, dass WIEGAND Spindeln mit unvollkommen fixierten, verklumpten Chromatinelementen vor sich hatte, die mir auch hier und da entgegentraten. Wass die niedrige Chromosomenzahl (6) anbetrifft, die WIEGAND für die vegetativen Kerne angibt, so muss ich auch die Richtigkeit dieser

Angabe in Frage ziehen. Bei allen meinen Zählungen konnte ich immer bestimmt das Vorhandensein von mehr als zehn konstatieren".

KOERNICKE does not give illustrations in his article; but, anyhow, it is clear that he saw eight gemini in the development of embryosac and pollen, whereas he saw more than ten chromosomes in the vegetative cells. Neither with this research was the number of chromosomes in *Canna indica* stated with absolute certainty.

BAUR in the first edition of his book „Einführung in die experimentelle Vererbungslehre" (1911) takes *Canna* as an example for the mechanical interpretation of Mendelian heredity under recognition of its having three chromosomes in gametic cells.

GRÉGOIRE (1912) mentioned *Canna sp.* with 2n=6 chromosomes.

HONING published in 1915 „Kreuzungsversuche mit Canna-varietäten". Obviously his intention was „zu prüfen, ob die Mendelspaltung auf der Verteilung der väterlichen und der mütterlichen Chromosomen bei der Reduktionsteilung beruhen kann oder nicht. *Canna indica* sollte nämlich generativ nur drei Chromosomen besitzen". The genetic research however proved that so small a number was not very probable. That is why HONING refers to the statement of KOERNICKE that *Canna* should have eight chromosomes in the generative cells. Through countings in vegetative cells he found sixteen as the diploid number. Later on (Canna crosses I, 1923) HONING mentions that he could confirm for his strain of *Canna indica* the number of chromosomes found by BELLING as haploid 9.

At that time several investigators tried to obtain a more accurate knowledge of the relation between chromosomes and the external characters of an organism. TOKUGAWA and KUWADA published an investigation in 1924 which was made already before 1919; it is a study of diploid and triploid garden-varieties of *Canna*. The diversity of the statements of WIEGAND and KOERNICKE induced the writers to re-investigate the number of chromosomes of *Canna*; moreover they made a comparative study in somatic characters between the diploid and triploid forms. (For the names of these varieties, see: TOKUGAWA and KUWADA, 1924). The numbers of chromosomes (investigated in the root-tips) are 18 and 27 (var. *Halley Comet* only with 17—18); 6 or 16 could never be found. (They gave 16 figures of somatic metaphase plates: 11 with 18 chromosomes and 5 with 27 chromosomes; somatic pairing or clear constrictions are not to be seen). „The number of chromosomes in the meiotic nuclear division was examined in several garden varieties. In *Eldorado, Halley Comet* and *Black Warrior* 9 gemini were counted with accuracy, as was to be expected from the number of chromosomes in the root-tips". In the triploid varieties the expected number of gemini and unpaired chromosomes was also found; the authors give a figure of an anaphase-spindle with 27 single or unpaired chromosomes scattered in the karyoplasm. „The process of meiosis in *Canna* is generally of somewhat abnormal tendency and in some cases it was almost impossible to count the exact number of chromosomes. The synapsis takes place as usual, but the linin substance seems to be rather poor in quantity. In the next stages the meagerness of the linin substance becomes more apparent. We can find only lines of chromatin-beads, which remind us of nuclear threads, but the connection between the beads is not so rigid as is usually the case in the latter. In the diakinesis the chromosomes are very loose in construction, sometimes so much so that one chromosome might easily be mistaken for two or even more. In such cases it is almost impossible to count the exact number of chromosomes or gemini. In *Eldorado, Halley Comet* and *Black Warrior* the behaviour of the chromosomes was somewhat regular, which enabled the writers to count the exact number as already mentioned. The division does not always result in only two daughter nuclei, as is usually the case, but very often three or more nuclei of various sizes can be found and many chromosome-like bodies with a clear circular space around each body. Thus we find more than one nucleus in a daughter cell. In these cases the nuclei may often appear as a larger nucleus of irregular form, connected with each other by a slender nuclear bridge a figure which reminds us of amitosis. The homotypic division is not yet fully studied, but can by no means be normal. So we have two or even more nuclei in a pollengrain, and not infrequently also chromosome-like bodies in the cytoplasm. The writers actually found a pollen-grain having five nuclei. In case two or more nuclei are found in a daughter-cell produced by the first division, they behave quite independently of each other in the second division, making the spindle in any direction as the chance may be. From this irregularity a disorder in the arrangement of cells produced from a pollen mother cell follows". „The cell division is abnormal, especially in the second division. In the first division we not infrequently find a newly formed boundary wall appear in section, not as a straight line as is usually the case, but undulated. In the second division abnormalities occur in a higher degree. In *Canna,* in the majority of cases the wall produced in the second division makes an acute angle to the first wall instead of a right angle as is normally the case in *Monocotyledons* and sometimes it appears even parallel to the latter. These abnormalities seem to be, though to a different extent in the different varieties, the usual case in *Canna,* either diploidal or triploidal. Thinking the abnormalities in the meiosis of the pollen mother cells might have caused the sterility of the plants at least in part, the writers examined the pollengrains of some garden varieties, and found in each case a certain amount of sterile pollengrains. Some garden-varieties produce seeds. Only a few of them barely germinate, but most of them do not come through. In the seeds examined we found most of the embryos degenerated. This, perhaps, has some relation to the abnormal combination of chromosomes due to the irregularities in the meiosis of the pollen mother cells".

It is remarkable to state how little has been said with certainty about the course of the prophase in former researches; it is probably a consequence of the fact that the earlier stages of development are difficult to investigate. Neither do the above authors say much about this matter and only give one drawing of a synaptic contraction and of a later stage of the var. *„Meteor"*. Probably it is a figure of early diplotene; the large paired granules are evident but the connecting threads are not shown in the picture. In all probability the authors could not see the connections, which is not very astonishing for the observation is difficult, because the achromatic substance of the chromosomes in *Canna* is nearly invisible in some stages, so that the chromatin-granules seem to lie separately in the nucleus.

In 1921 JOHN BELLING's article appears: „The behaviour of homologous chromosomes in a triploid *Canna*". In the preceding years much cytological work has been done and the interest is directed to the pairing of the chromosomes in prophase and metaphase and to the separation in the anaphase; it is to be expected that researches of polyploid forms will give a better comprehension of these phenomena about which there are still so many different opinions. BELLING says: „In 1920 I grew 46 differently named clones and species of *Canna*, obtained, without particular selection, from three leading dealers of New-York, Philadelphia and Florida and from the U.S. Department of Agriculture. The chromosomes of 31 of these were counted in the first or second divisions of the pollen mother cells; about 250 groups being drawn with the camera. Of these 31 clones, 22 showed nine dyads or bivalents which split into 9+9 at the first division; 3 clones had a total of 18 single chromosomes (or 9 bivalents), which divided at the first division into 8+10, or other unequal numbers, not commonly into 9+9; 5 clones were probably completely or nearly triploid, and irregular in their first division and .... showing a smaller number of chromosomes after the first division than the triple number (in these cases 24 to 26 instead of 27); while one clone was regularly triploid, showing nine triads at the prophase and first metaphase, and a total of 27 chromosomes after the first division. Thirty-two pollen mother cells were drawn with the camera. In 18 of these the total number of chromosomes could be accurately counted, and was 27''. This regularly triploid clone was obtained from Thornburn, New-York in 1920 under the name „*Gladiator*''. It is a smal form as the diploid Cannas. (In the paper of TOKUGAWA and KUWADA, „*Gladiator*'' is a diploid form! The investigators gave no further information in the text neither pictures).

In his summary BELLING states: *a*) „Most of the Cannas examined were diploid, showing nine dyads, before the first division in the pollen-mother-cells, and these in most plants separated into 9+9. *b*) One of the triploid Cannas showed commonly nine triads, each of which separated into two and one on the spindle, in a random manner with regard to the two poles''.

The already mentioned researches of TOKUGAWA and KUWADA of 1924 have been published in consequence of BELLING's paper of 1921.

In 1925 BELLING publishes again about *Canna*. (BELLING 1925*a* and 1925*b*). In the first paper he mentions *Canna* only as an example of a genus with triploid forms and he observes, that „the chief characteristic of true triploids is a partial or total sterility; the triploid *Cannas*, for instance, never (or rarely) producing seeds''. „The characteristic of the tetraploids is the possession of much larger cells''. The production of a triploid apparently originates with the doubling of the chromosome number in one of the gametes. This doubling is in many cases certainly due to a temporary chill, resulting in non-reduction or non-division during the maturation divisions of pollen mother cells, or megaspore mother cells (or to non-division in the somatic mitoses which give rise to the sub-epidermal tissue of a branch). Triploid plants are already of commercial value for their flowers in the case of a few Cannas". BELLING says further, that the production of triploids should be tried by exposing the young flower buds to low temperatures.

The second article gives the figures, belonging to BELLING's publication of 1921. The preparations were made with the iron-aceto -carmine method. The pictures show:

a) Metaphase I in a pollen mother cell of the ornamental Canna „*Gladiator*" with 9 trivalents. The connections between the chromosomes of each trivalent are different; they can form a chain, a V or a tri-radial. Their finer structure is not visible in the pictures.

b) The same stage of the Burpee Canna „*Firebird*", which clearly shows that 27 chromosomes form 9 trivalents.

c) The reduction division in the Canna „*Pennsylvania*", which shows trivalents, bivalents and univalents.

Meanwhile the aceto -carmine method was put into practice more and more for preliminary investigations, especially for countings of chromosomes. HEITZ publishes in 1926 an article about the application of this method, in consequence of the results of an investigation. „Ich untersuchte *Canna flaccida* und zwei Gartenvarietäten von *C. indica,* ebenfalls Wurzelspitzen. Für die eine könnte gleich im ersten Präparate die Zahl 18 mit Sicherheit festgestellt werden, für die zweite mit groszer Wahrscheinlichkeit. Es liess sich nicht entscheiden, ob 2n = 18 oder 20 ist. Dagegen zeigten die Platten der noch nicht untersuchten Art *C. flaccida* einwandfrei ebenfalls 18 chromosomen".

HONING gives in 1928 a review of his new investigations of pollen-mother cell divisions. He determined the chromosome number of *Canna glauca* and *aureo-vittata,* which is haploid 9. For the greater part his article deals with the phenomena in the reduction-divisions of *Canna aureo-vittata gigas* which was obtained by crossing *Canna aureo-vittata deep yellow* × *pale yellow*. This new type is a real gigas form, as well from the genetical point of view as from the cytological one, for it has, either *a*) the double number of chromosomes or *b*) nine very big (tetravalent) chromosomes (in the metaphase I) or *c*) in other pollen mother cells uni-, bi- and tetravalents mixed. The *C. glauca*×*indica* F_1 hybrid also often shows univalents in diakinesis. In a few cases HONING found chains of chromosomes with different forms. He leaves undecided the method of pairing of the chromosomes in early prophase, either para- or telosyndesis.

KRACAUER (1930, „Die haploidgeneration von *Canna indica*") describes the development of pollen and embryosac. With regard to the meiosis of the pollen mother cells he says: „Die heterotypische Teilung des Kernes geht in ihren Phasen völlig regelmäszig von statten". This does not correspond with the words of TOKUGAWA and KUWADA, quoted above, but KRACAUER means probably the course of the development in general and not details of the prophase. He evidently investigates a regular strain, with n=9 chromosomes: „Während aller Teilungs-phasen sind die Chromosomen in der Haploidzahl (neun) mit Sicherheit zu zahlen". It is a pity that KRACAUER does not give details about the prophase, for according to his own words the development of the embryosac mother cells proceeds slowly from the one stage to the other.

The figures of his paper are too small to give a good idea of the structure of the nucleus contents. KRACAUER is of opinion that in the embryosac mother cells a continuous spireme has developed, which splits lengthwise in the synapsis (synezisis). He observes that the gemini in diakinesis are not regularly scattered through the nuclear cavity and that they do not form chains in this stage.

V. MATERIAL AND METHODS

Observations are chiefly concerned with the reduction divisions of the pollen mother cells; sometimes somatic tissue of the anthers afforded good opportunities for counting chromosomes.

The inflorescenses of *Canna* are panicles (composed of cincinni) which develop terminal and in the axils of the higher leaves of the floriferous stalks; as a rule two flowers are close together, sometimes with a small rudimentary third flower. If a shoot is already too far developed, flowers containing the desired stages of the pollen mother cells are often found in the secondary one which is still inside the leaf.

The inflorescences were cut off and immediately put into water; the stamens were prepared and fixed in the laboratory as soon as possible. The Canna flower (fig. 2) has only one half stamen fertile (fig. 3). In order to avoid the collection of unsuitable material, a preliminary inspection of the contents of a little piece of each anther

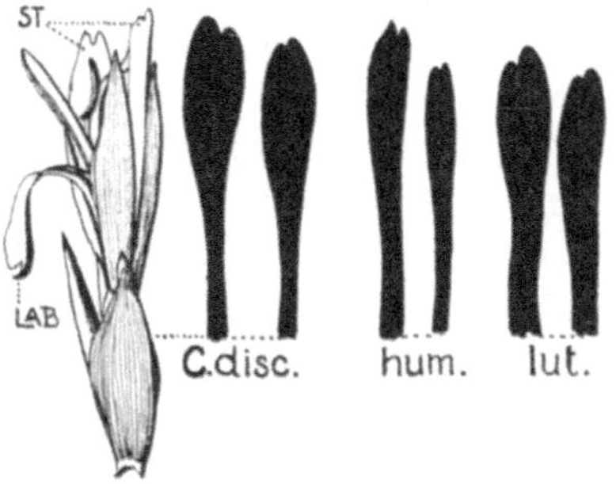

FIG. 2. Flower of *Canna discolor*; staminodes of *C. discolor, humilis* and *lutea.*

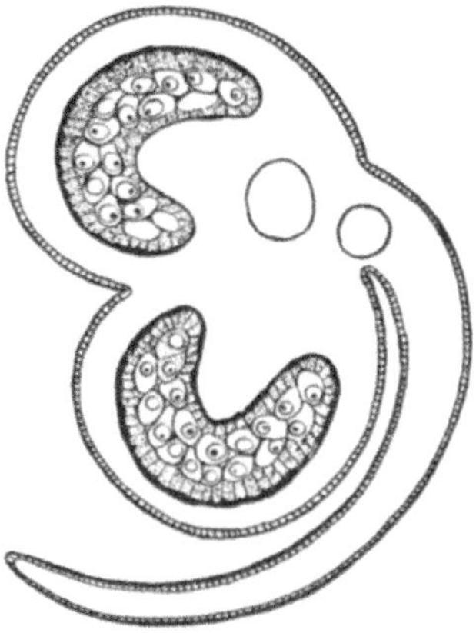

FIG. 3. Anther, diagram of a transverse section.

was made in aceto-carmine (in accordance with the indications of BELLING, 1921, 1926, 1928, 1930, HEITZ, 1926 and DE MEIJERE, 1928, 1930).

Later on in order to be able to state exactly the transition of the different stages in the preparations, young complete inflorescences were fixed in the greenhouse. I kept the anthers under the fixative at the bottom of the tube by means of a dot of cottonwool; this turned out to be not only a simple method but also a good one. The superfluous parts of the flowers were removed later on as much as possible in the course of the dehydration. These complete inflorescences were put into paraffin.

The following fixatives were chosen: CARNOY, FLEMMING (strong mixture), BOUIN-ALLEN and modified NAWASCHIN (cf. LELIVELD, 1928).

Bad results were obtained with strong FLEMMING, because of considerable shrinkage and distortion.

CARNOY's fluid is suitable for all stages; it may be an excellent one for the prophase.

BOUIN-ALLEN's fluid is generally a bad fixative for the leptophase, good for diakinesis, meta- and anaphase.

NAWASCHIN's fluid (modified) is fairly good for all stages, but like BOUIN-ALLEN's fluid, it gives a coarse precipitation of the nucleus-contents in the early prophase. The chromosomes in meta- and anaphase are more swollen and lumped together, which reduces the possibility for countings and further study.

The imbedding in paraffin wax was first done in the usual manner. Later on I proceeded to a modified method which in general may be compared with those of BAKER (1933).

As a rule the blocks were cut 18μ thick; very young inflorescences sometimes 15μ and to increase the possibility of getting sound nuclei sections of 20μ were also made. The greater part of the material was stained with HEIDENHAIN's iron-haematoxylin, which gives an exceptionally clear staining of the chromosomes and their chromomeres. The staining with NEWTON's gentian-violet gave satisfaction for the chromosomes in meta- and anaphase, but not for the earlier prophase stages. Comparing the stainings (on slides with sections from the same anthers and with the same stages) the haematoxylin is certainly preferable.

All drawings were made with the aid of a drawing-apparatus (at the level of the work-table); objectives Zeiss H.I. 90× and 120×, Fluorit 100× with the compensating oculars: 10×, 15× and 20×.

VI. DESCRIPTION OF THE MEIOSIS OF THE POLLEN MOTHER CELLS

A. *Canna humilis* BOUCHÉ

The following description of the development of the pollen mother cells starts at the moment when they are clearly to be distinghuished from their surroundings, i.e. when the differentiation of tapetum and archesporium has been accomplished. When this stage has been reached, tapetal cells and pollen mother cells are conspicuous by retaining stain in a higher degree than the cells of the surrounding layers. The transverse section of an anther shows two locules with an epidermis of one layer of square cells with large nuclei; beneath that about seven layers of parenchyma-cells follows, of which the outermost rows show a more rectangular shape. (See for a metaphase plate of these cells fig. 4). They become gradually smaller in radial diameter in the direction of the inside. The innermost tapetal cells already very soon contain more nuclei. Here and there in this stage the (mitotic) division of the tapetal cells can be seen. Sometimes very large tapetal cells are to be found with a great number of chromosomes, probably originated by imperfect division. The development of the tapetum and the degeneration of the tapetal cells are no safe indications for the stage of developing of the pollen mother cells themselves, for sometimes the changes proceed very slowly, in other cases very rapidly.

FIG. 4. *Canna humilis*, Somatic metaph. plate, Carnoy, 3600 ×.

The archesporium (fig. 3) generally consists of two rows of large multangular cells, the pollen mother cells. As a rule one is as far developed as another in the same loculus; if any difference is to be seen in development at all, then generally the pollen mother cells in the upper part are in advance of the others. With fixation in Bouin-Allen's fluid we get the following impression: The pollen mother cells have very large nuclei in which a great dark-coloured nucleolus is immediately prominent; usually the latter lies some way off the nucleuswall and somewhat excentrically. The greater part of the nucleus-hole is clear; it is filled with a fine reticulum of faintly stained chromatin-threads. The more precise structure of these threads which are not quite straight could not be accurately examined. In the nuclear cavity a number of small round bodies are to be seen, which have assumed the same colour as the nucleolus; regularity in the appearance of these small bodies, as regards size or number could not be observed (fig. 5).

The granular substance of the reticulum becomes gradually more clearly visible, the granules become more darkly coloured and larger; the threads themselves are very thin, sometimes hardly visible. Later on they are directed more meridional and often run parallel, but it is not possible to see a pairing in this phenomenon. Free ends in sound nuclei are not to be seen (fig. 6).

This stage is followed by a contraction of the reticulum with the coherent chromatic material in the vicinity of the nucleolus: obviously this does not happen suddenly, but gradually, for transitional stages are to be found in the preparations. (The typical synaptic knot is an artefact, undoubtedly, owing to bad fixation).

Further advanced stages are to be found in the same preparations: in the nuclear cavity a clear thread is to be seen, lying in large loops. It is impossible to say whether we must speak of a continuous spireme

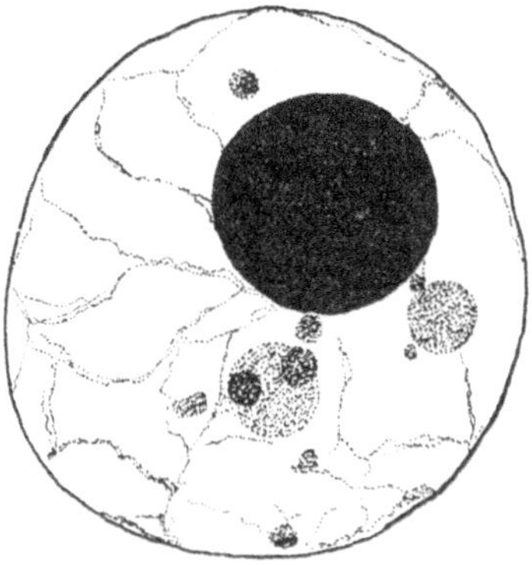

FIG. 5. *C. humilis*, Nawaschin, early-zygophase, 1800 ×.

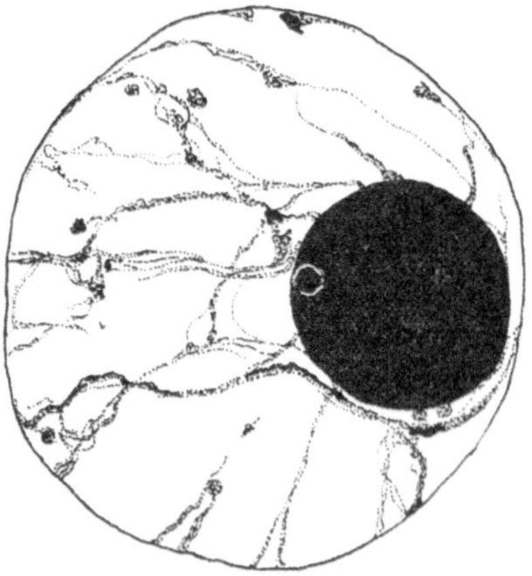

FIG. 6. *C. humilis*, Bouin-Allen, mid-zygophase, 1800 ×.

or of separate loops; free ends are not to be seen. At a later stage the threads contract; their smooth sides disappear and they become thicker and get a more granular surface. Sometimes the threads show signs of doubleness in that their margins are darker and the centre lighter.

As it appears from the following explanation the stage just mentioned is zygotene-pachytene. On comparing preparations, which were made with different fixation-fluids, it appeared that the fixation with Bouin-Allen's or Nawaschin's fluid gave an incorrect idea of the structure of the nuclear contents in the early prophase stages. Both fixation-fluids mentioned, as a rule, give too coarse a precipitation, so that the finer structure, namely of leptotene-threads and of the chromomeres is totally absent. On account of this the actual pairing of the leptotene-threads and the exact development of the zygotene-stage becomes invisible. It is not superfluous to point out once more that the aspect of the chromosomes may be to a great extent dependent on the fixative used; different fixatives give rise to different figures in the same stage.

This again may be the explanation for the fact, that several researchers have altogether different opinions about the development in the same material, especially with regard to problems, such as those of the para- and telosyndesis. Now in those cases, where good results were obtained with *Canna humilis* when fixed in Carnoy's fixation-fluid, the leptotene-stage presented no difficulties in this respect, so that the development of the bivalents could easily be traced. Moreover they were of importance for the explanation of the figures of the chromosomes, obtained by using the other fixatives mentioned. It is conspicuous that the nucleolus is not so dark

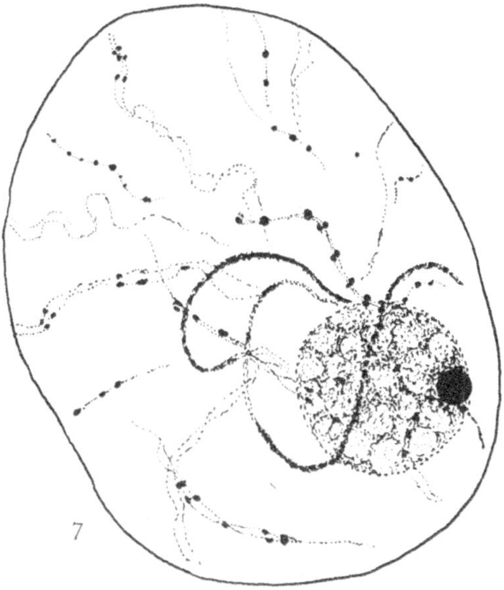

7

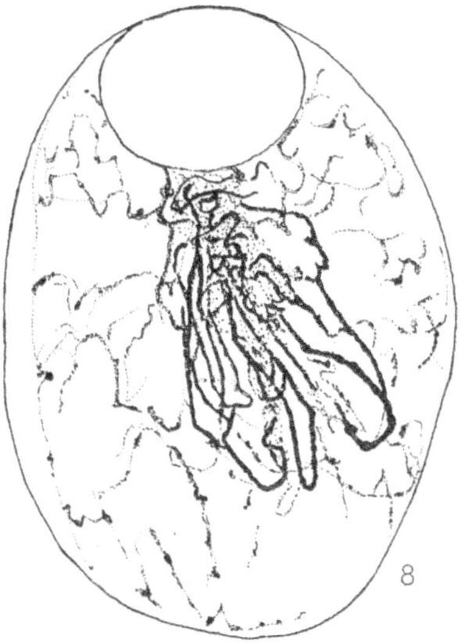

8

FIG. 7-8. *C. humilis*, Carnoy, zygophase, 3600 × and ± 3000 ×.

coloured as after using Bouin or Nawaschin; it seems to be a globular body of alveolar construction. Similar large chromatic bodies are not present (with the exception of one small round body in the immediate neighbourhood of the nucleolus).

By conscientiously studying this stage it can be concluded that the very fine threads associate in pairs; in favourable circumstances the chromomeres of two threads are seen to be lying in corresponding places (homologous threads with homologous chromomeres, fig. 7). The pairing threads contract, lengthwise, on account of which the chromomeres approach each other more and more, for the developing thicker thread is smooth and dark-coloured. Usually leptotene-threads and zygotene-threads are found to be lying together in a

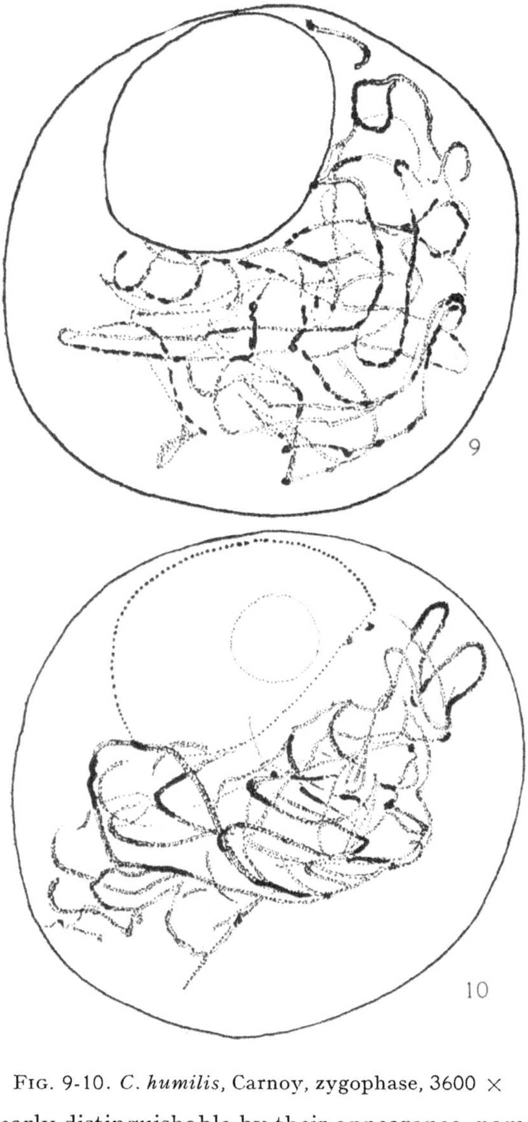

FIG. 9-10. *C. humilis*, Carnoy, zygophase, 3600 ×

nucleus; clearly distinguishable by their appearance, namely *a*) thin threads with fine chromomeres at unequal and fairly large distances, and *b*) thick, dark-coloured threads with chromomeres united into a continuous row (fig. 8). This gives the so-called amphitene of JANSSENS. This stage passes into the ultimate zygotene; it produces the aspect of the „open spireme", large loops lie throughout the nuclear cavity (fig. 9 and fig. 10). In early zygotene the threads show light-coloured or nearly colourless spots (fig. 11); it is not impossible that these „gaps" become the places for the spindle-fibre-attachments (attachment-constrictions), but their further development was untraceable here. Obviously the threads of this „open spireme" contract. (Whether this is a „continuous" spireme or not, cannot be definitely stated; free ends are not to be seen.)

They grow thicker and often come to lie closer together; their sides become corrugated, i.e. more clearly granular. In the middle of these threads a lengthwise split becomes visible; the double structure of the threads is apparent. What has happened to the chromomeres or the chromatids, cannot be perceived. It cannot be stated with certainty that in this stage four chromatids constitute the thread. The pachytene-loops expand and fairly suddenly the nuclear cavity is filled up with the diplotene threads (fig. 13). Now the chromomeres are visible again, situated on twisted and crossed paired threads. At this stage it is not possible to study the whole complement, for the chromosomes are so long, that they cannot be individually traced; it stands to reason that the connections of the chromatids, the chiasmata, are not visible either. If the crossings and twists of the threads are to be considered as the points where the chiasmata are formed, it is probable that these are formed at random. The precise number of chiasmata is not to be ascertained (but there may be five or six to each bivalent). This stage changes into diakinesis in a remarkable manner for the contraction of the bivalents does not take place along their whole length simultaneously. The contraction of the diakinesis-bivalents is gradual; when local condensation has already materially advanced, the remainder is still in the diplotene stage (fig. 12*a*, 12*b* and 14 *a*—*i*). At the same time both the diplotene and the diakinesis-stage are observable, namely, the various bivalents of the same complement do not contract synchronously which is neither the case with the development of certain portions of each bivalent. This remarkable phenomenon might be called heterochronous condensation of the bivalents (fig. 15*a*, 15*b*). Although it cannot be ascertained accurately how many chiasmata were originally formed, it is

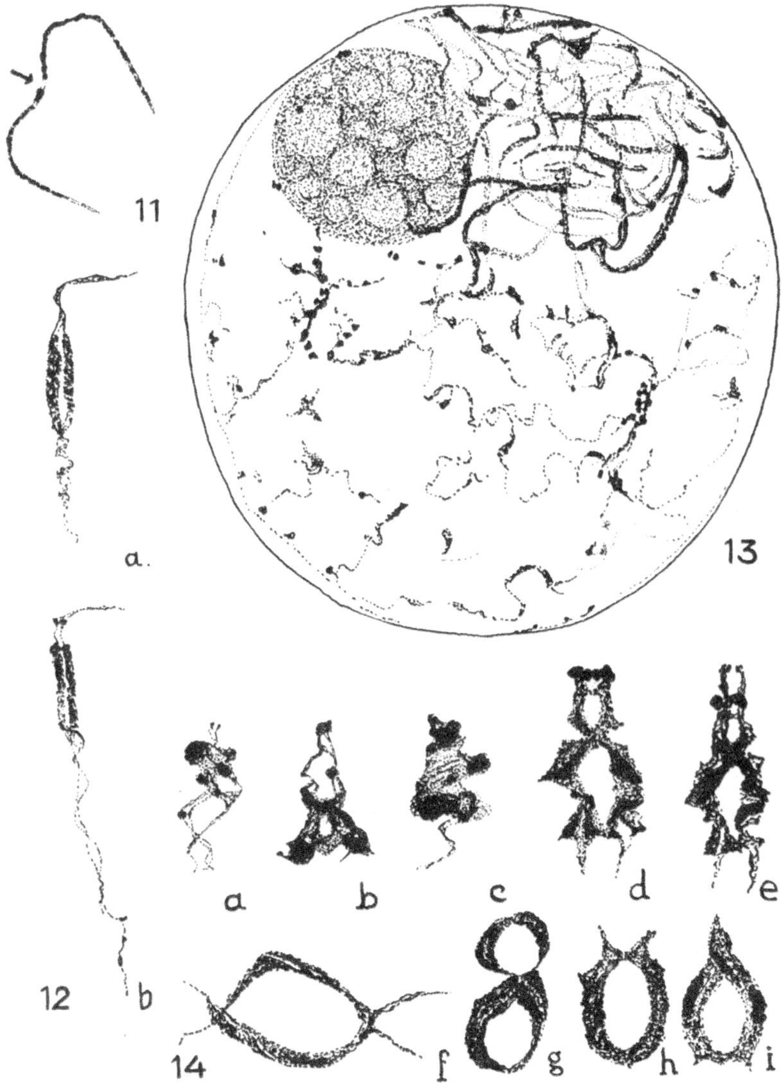

FIG. 11. *C. humilis*, Carnoy, „spindle-attachment" constriction, 3600 ×.
FIG. 12. *C. humilis*, Bivalents (partly condensed), 3600 ×.
FIG. 13. *C. humilis*, Transition pachyphase-diplophase, 3600 ×.
FIG. 14. *C. humilis*, Bivalents, diplophase, a, b, c, 3600 × ; d-i, ± 4000 ×.

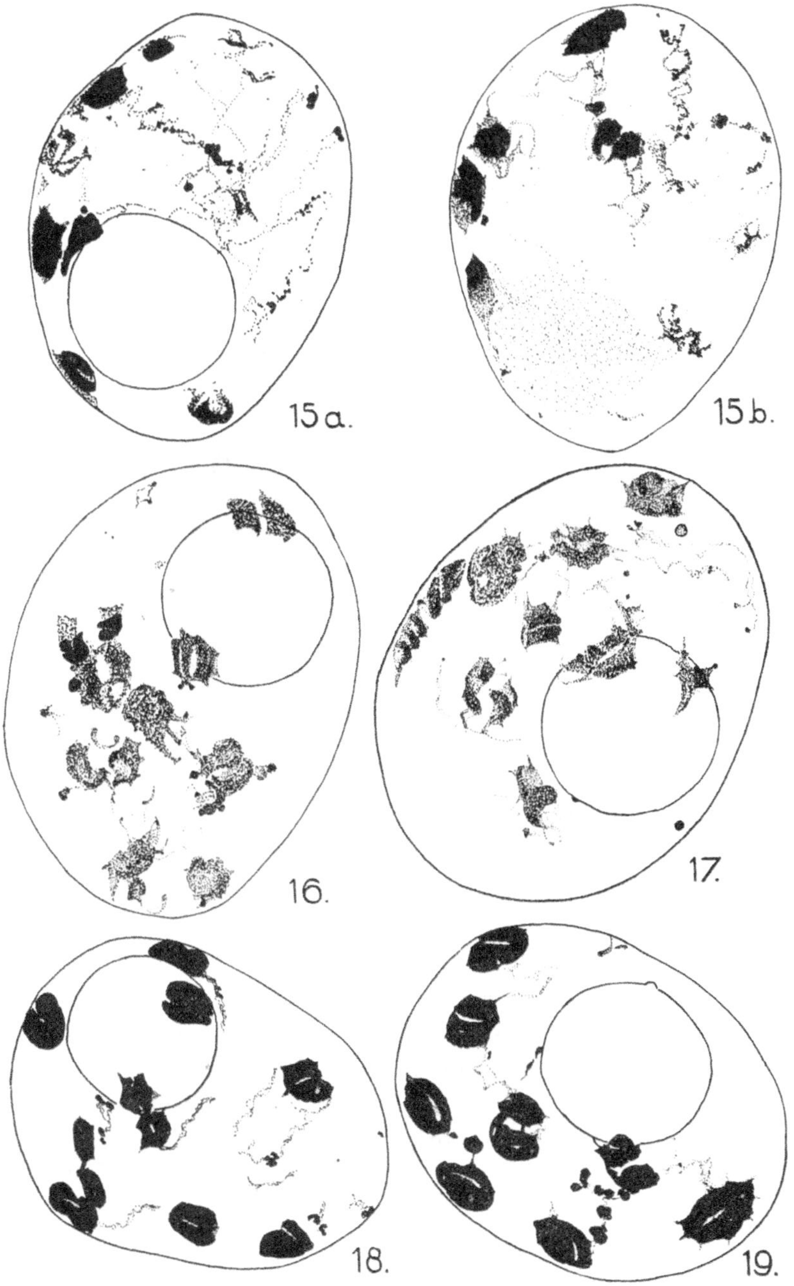

FIG. 15. *C. humilis*, Carnoy, diplophase, two fig. of the same nucleus, 2400 ×.
FIG. 16-19. *C. humilis*, Diakinesis-development, nine bivalents, 2400 ×.

clear, that their number gradually diminishes (terminalization); see fig. 16, 17, 18 and 19. Finally the chiasmata are all terminal. At the end of diakinesis the condition of the bivalents is nearly uniform, for most of them are ring-bivalents, except one or two, which are rod-bivalents (fig. 18). The „end-to-end association" of these bivalents has arisen owing to the „terminalization" (probably) of the chiasmata. When pursuing the development of the bivalents in diplotene, the impression is created that the chromatids in each

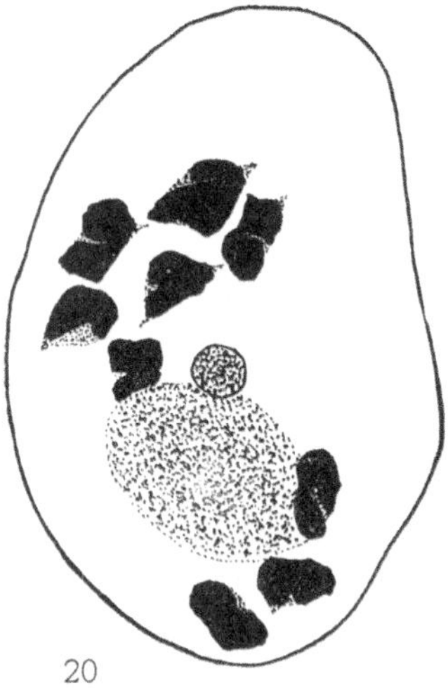

20

21

FIG. 20. *C. humilis*, Diakinesis, 3600 ×.
FIG. 21. *C. humilis*, Bipolar nuclear cavity; fine threads visible with small stained particles, ± 3600 ×.

pair are spirally twisted (fig. 12, 14-17). They contract, the spirals become shorter and thicker; finally both homologous parts of the bivalents give the impression of a double thread. Ultimately the bivalents are totally condensed into nine double bodies (fig. 20). Meanwhile the nucleolus decreases in size, whilst at the same time a smaller body of a similar appearance becomes visible in the immediate neighbourhood of it (probably there is some connection between them). When the nuclear cavity disappears, the spindle figure becomes visible (fig. 21). Possibly this does not take place instantaneously, though transitional stages are rarely to be seen. A multipolar spindle could not be observed. After having first been scattered in the spindle (fig. 22), the bivalents are drawn up in the centre. Sometimes they lie in a cluster (fig. 25) before they assume their position in the metaphase-plate (fig. 23). The metaphase I is very suitable for counts; in several preparations this stage is frequent and moreover the bivalents are clearly distinguishable in polar views.

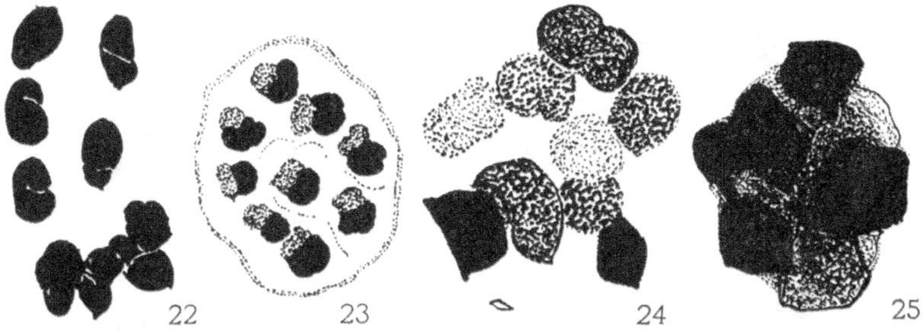

FIG. 22. *C. humilis*, Nine bivalents scattered through the spindle, 3600 ×.
FIG. 23. *C. humilis*, Metaphase I, 3600 ×.
FIG. 24-25. *C. humilis*, Nawaschin, chromosomes swollen, 3600 ×.

They are equal as to form and size. In anaphase I the individuals of the bivalents separate; this is a quick process (fig. 26). The transition from metaphase I to anaphase I is in case of *Canna humilis* not very suitable for observation of the remaining chiasmata.

From the diakinesis the conclusion may be drawn that a complete terminalization is brought about (or, at any rate, is probably brought about), but in early anaphase I the chromosomes are too small and too compact, to confirm the above statement. Separation of the chromosomes in anaphase I is not always synchronous; one or two may be somewhat in advance of the others. They may possibly have arisen from the rod -pairs. This too cannot be proved with absolute certainty, as the closely associated bivalents in metaphase I are not distinguishable (fig. 23). Neither is there any striking difference between the chromosomes of an anaphase-complement (fig. 26*a*, *b*). A great number of corresponding anaphase-plates are found in the preparations; the separation in 9+9 is very regular (fig. 26, *c*—*f*). In clear cases the anaphase chromosomes are seen to be double; their aspect changes from early to late anaphase (fig. 27). At first four free extremities are visible, connected at one point at least; later on the ends are drawn together and a lengthwise split is conspicuous. This

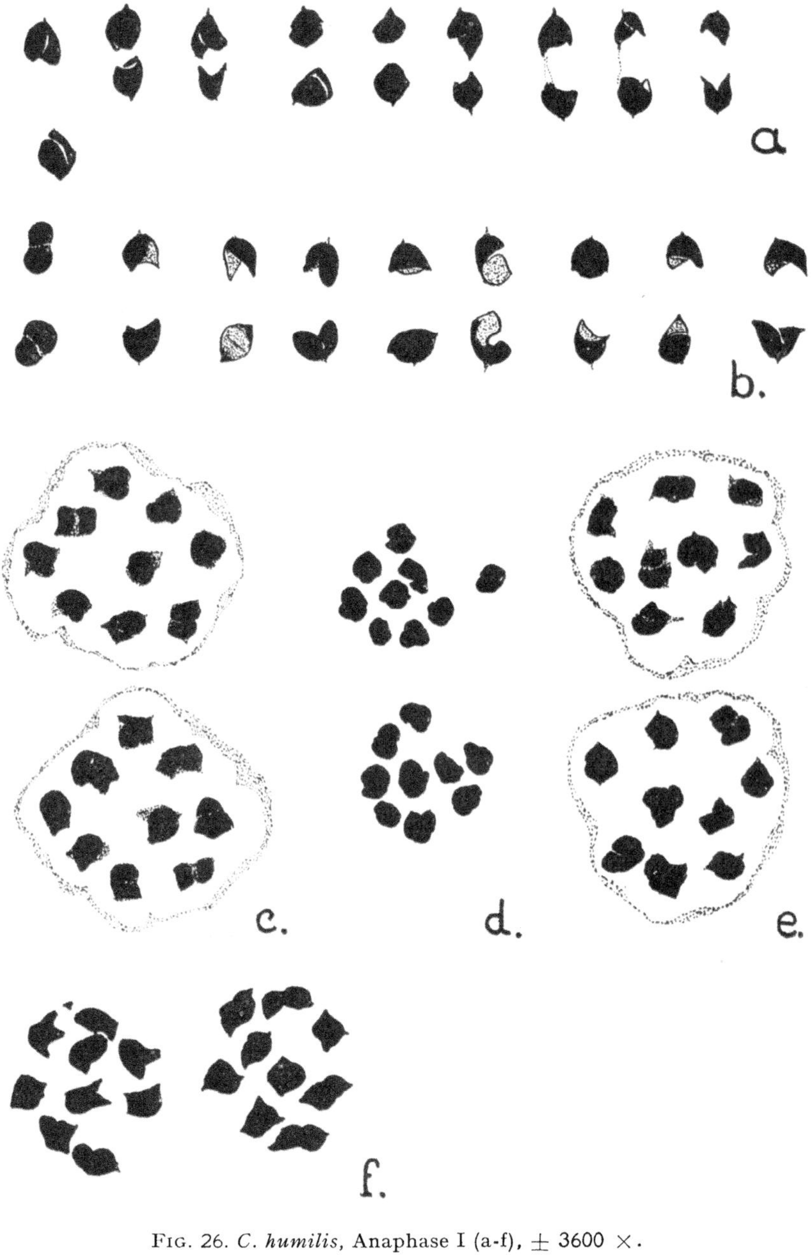

FIG. 26. *C. humilis*, Anaphase I (a-f), ± 3600 ×.

is still the case in telophase I; the chromatids which are connected at the ends and at a point towards the middle are clearly distinguish-

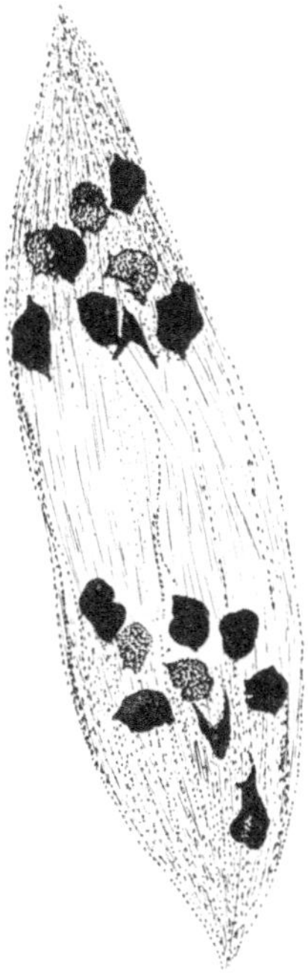

FIG. 27. *C. humilis*, Anaphase I, 2400 ×.

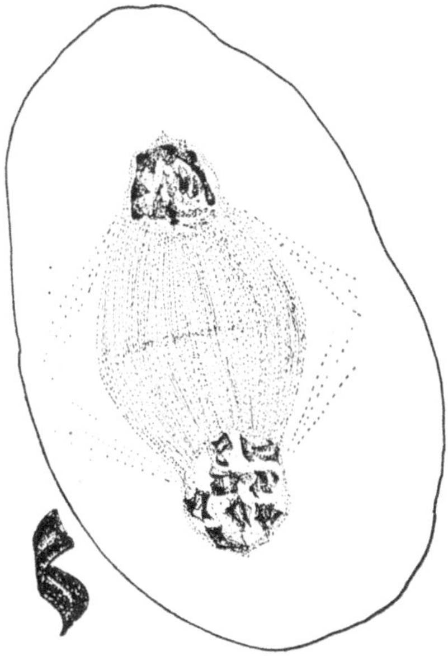

FIG. 28. *C. humilis*, Telophase I, 1200 ×.

able; this gives the impression of an eight-shaped figure, which is slightly spiral-formed (fig. 28).

The spindle has in the meantime become more globular and has expanded in the width (phragmoplast); in the middle arises a new cell-wall, separating the two cells of the dyad. This takes place as usually in monocotyledonous plants (fig. 29—33). Occasionally nine chromosomes may still be easily counted in the nuclei of the dyads (fig. 30); they are often constricted in the middle. As a rule vacuolization soon appears, so that the individual chromosomes are no longer distinguishable. Little can be said about the duration of the interphase; sometimes this stage is exclusively found in both the loculi of an anther, at other times dyads and tetrads are found together, with all the phases of the second meiotic division. In metaphase II, the long and narrow spindles (fig. 29) often lie parallel or nearly parallel with their long axes, but they may also cross each other. The factors, causing these different positions, are not to be inferred from the preparations.

Anaphase II proceeds in a regular manner; several times it may be

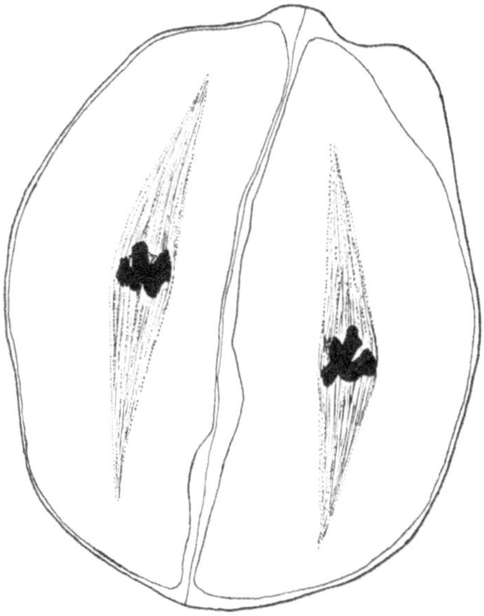

FIG. 29. *C. humilis*, Metaphase II, 1200 ×.

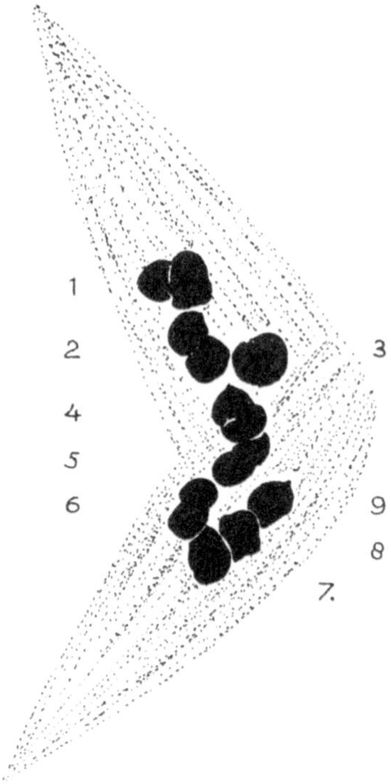

FIG. 34. *C. humilis*, Prophase; abnormal spindle, 3600 ×.

seen that the chromosomes separate regularly in two sets of 9 each (fig. 32). Very often, however, the fairly small chromosomes are grouped together, offering difficulties for observation. The tetrads arise in the usual manner (fig. 29, 33).

There are abnormalities (e.g. fig. 34, an abn. spindle-form), but in such a small percentage that they are of no significance for our considerations. The pollen is formed abundantly and only a very few pollen-grains are sterile.

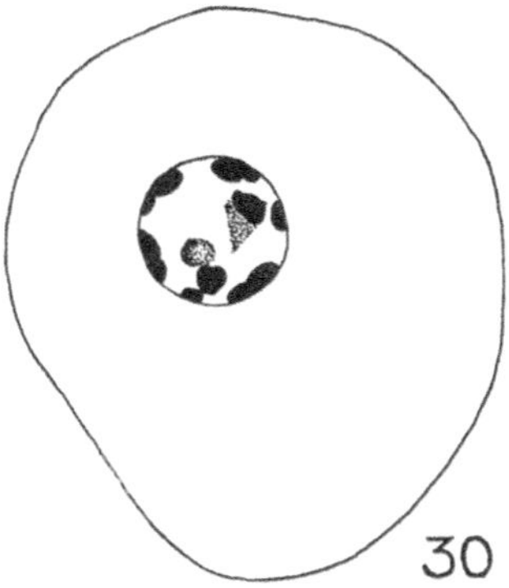
30

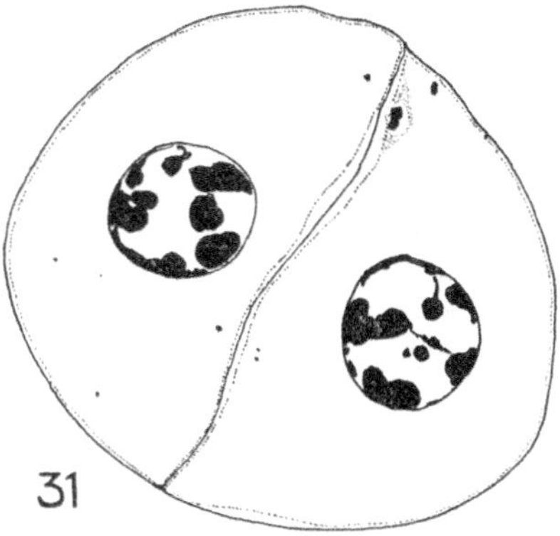
31

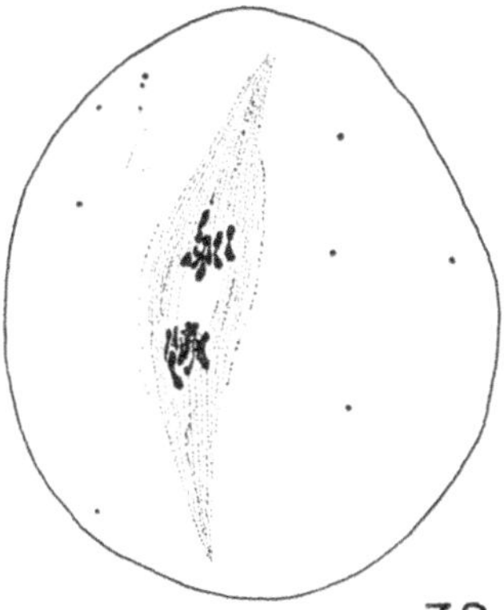
32

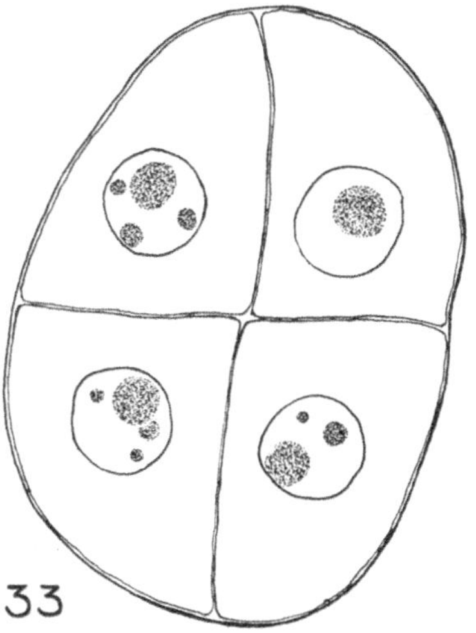
33

FIG. 30-31. *C. humilis*, Interphase, 900 ×.
FIG. 32. *C. humilis*, Anaphase II, 900 ×.
FIG. 33. *C. humilis*, Tetrade, 900 ×.

B. *Canna lutea* MILL.

The zygophase and pachyphase develop as described for *Canna humilis*. If differentiation is correct, in late pachyphase details are observable in the threads, which give the impression, that these really consist of four strings (fig. 35). This stage passes into diplotene (fig. 37); with this species too it is not possible to state with certainty the development of the chiasmata and their exact number, firstly

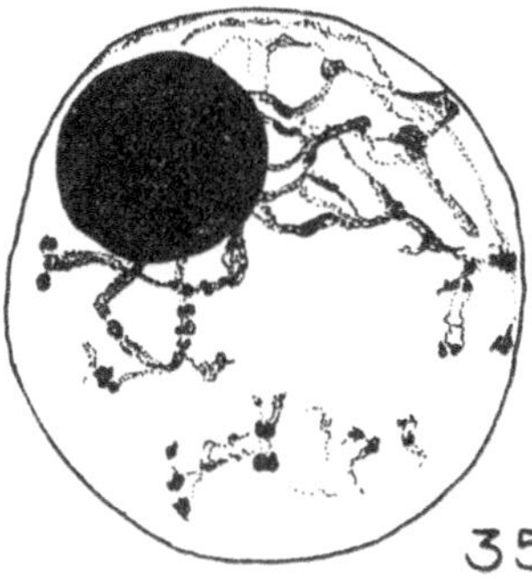

35

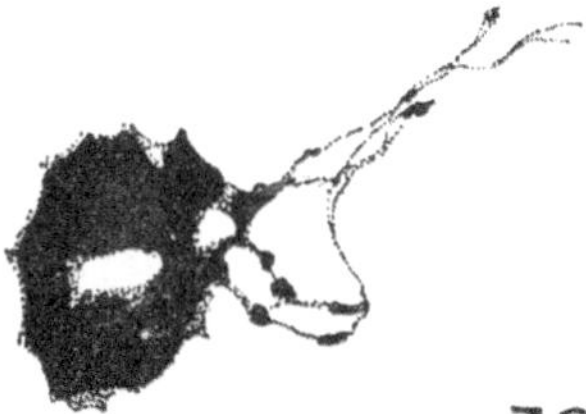

36

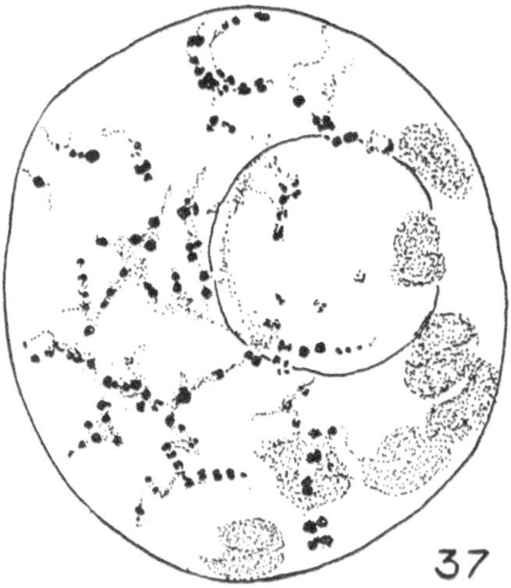

37

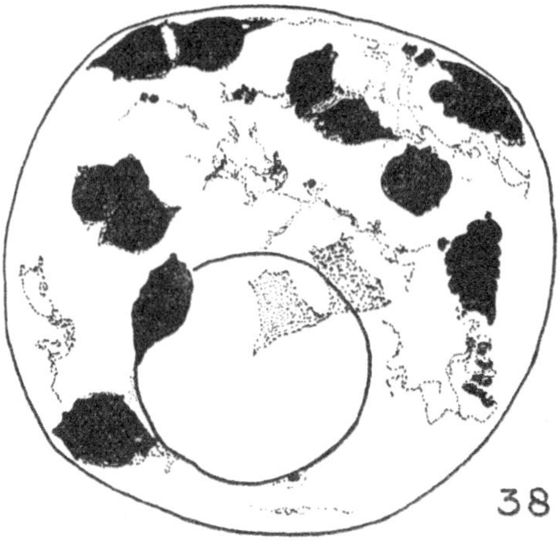

38

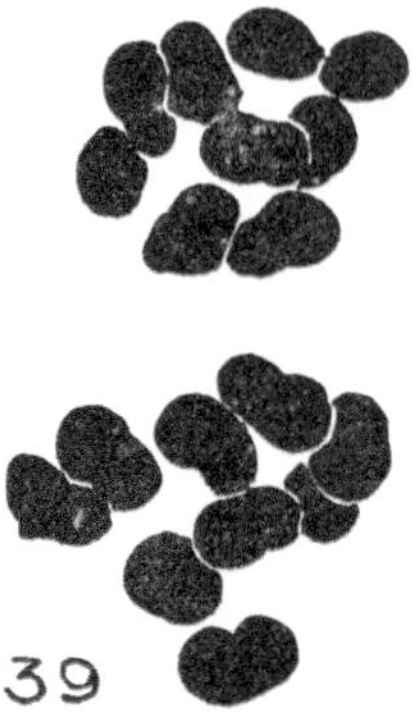

39

FIG. 35. *Canna lutea*, Pachyphase, 1350 ×.
FIG. 36. *C. lutea*, Bivalent, late diplophase, ± 3600 ×.
FIG. 37. *C. lutea*, Early diplophase, 3600 ×.
FIG. 38. *C. lutea*, Late diplophase, 3600 ×.
FIG. 39. *C. lutea*, Anaphase I, 3600 ×.

because the chromonemata are extremely hard to trace, secondly on account of the fact that the contraction in the bivalents locally is much further advanced than elsewhere (fig. 36 and 38). The whole process proceeds gradually, but not quite simultaneously for all the bivalents in the same nucleus. Finally 9 bivalents are developed, which contract more and more. The paired chromatids of a bivalent are obviously spirally coiled round each other. It is clear that only terminal chiasmata remain; a single rod-bivalent is developed.

The transition from prophase I to metaphase I etc., and the second meiotic division can be a quick process, for all these stages were found in a single anther. The process is, as a rule, very regular; in most of the cases observed, the chromosomes separate in equal number (9+9) to either pole (fig. 39). The dyads look very normal, as is the case with the tetrads. Sometimes there are deviations: *a*) divisions with lagging chromosomes and *b*) divisions, which give rise to unequal numbers. What may be the outcome of these abnormal divisions, remains to be seen. The abundantly formed pollen looks very normal; conspicuous abnormalities are certainly absent.

C. *Canna discolor* LINDL.

As already mentioned (in the introduction) the variety examined is presumably of hybrid origin; it has a regular development as to the meiosis of the pollen mother cells. The stages of the prophase are essentially similar to those of *Canna humilis* and the other form described. The leptotene-threads associate in pairs synaptically, (fig. 40); the transition from one stage to the other proceeds gradually, especially the condensation of the bivalents. With *Canna discolor* meiotic stages may be found where some of the chromosomes form chains, but it is a rare phenomenon; moreover it is very well possible that in most of the cases such a figure will be a „pseudo-chain" (only optically a chain) with the bivalents lying close together in a straight or in a curved line. In metaphase I connected chromosomes are not to be seen: nine bivalents are distinctly developed (fig. 41—42). The separation of the chromosomes in anaphase I proceeds regularly; the splitting of the anaphase-chromosomes is evident and their form changes from early to late anaphase, as already described for the other species (fig. 43).

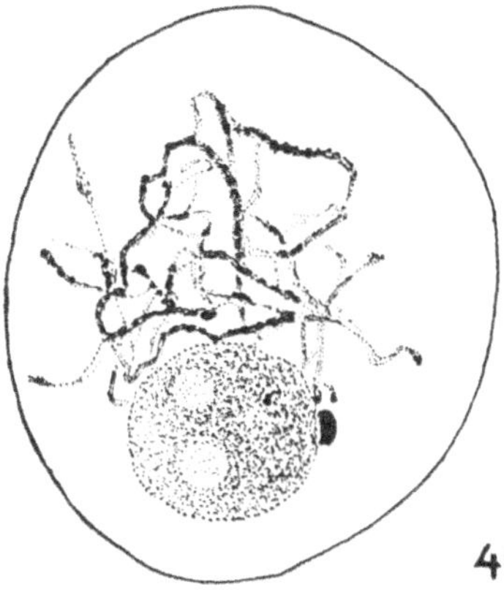

40

41

42

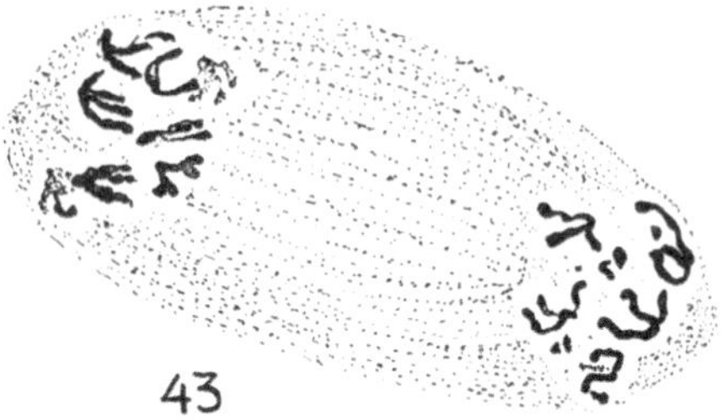

43

FIG. 40. *C. discolor*, Zygophase, 1800 ×.
FIG. 41. *C. discolor*, Anaphase I, 3600 ×.
FIG. 42. *C. discolor*, Metaphase-plate I, 3600 ×.
FIG. 43. *C. discolor*, Anaphase I, late, 1800 ×.

For the rest the former descriptions may be referred to.

D. *Canna glauca* L.

(The meiosis in the varieties „JAVA", „BOLIVIA", „MONTEVIDEO" and „PURE YELLOW".)

SOMATIC DIVISION: Frequently, somatic metaphase plates in the parenchyma cells of the wall of the anthers could be observed; countings resulted in 18 chromosomes (fig. 44). Their finer morphological structure could not be studied, for the fixation-fluids used, were not suitable for that purpose. Notwithstanding, in the same anther e.g. the nuclei of some tapetal cells can show very fine fixed chromosomes with clearly visible constrictions and trabants. Owing

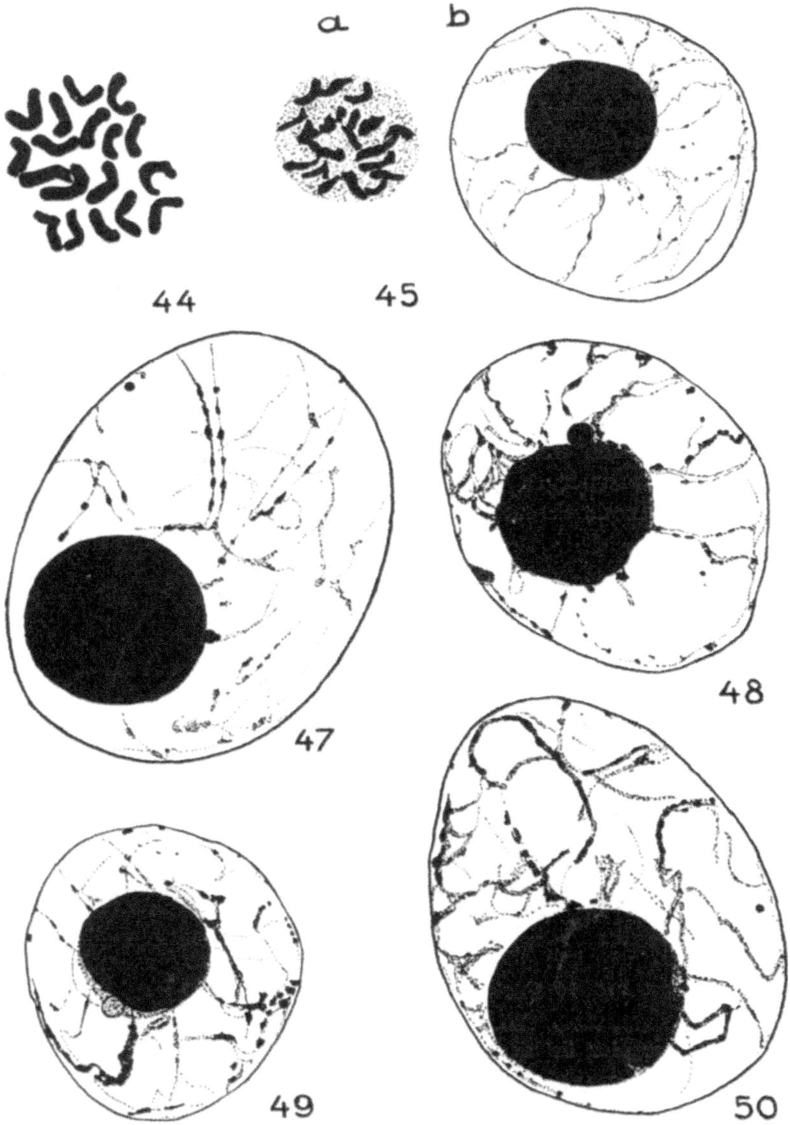

FIG. 44. *C. glauca* „Java'', Somatic metaphase, 3600 ×.
FIG. 45. *C. glauca* „Java'', *a*. Nucleus of a somatic cell, and *b*. Pollen mother cell at the same magnification, ± 2000 ×.
FIG. 47 *C. glauca* „Java'', Zygophase, 3600 ×.
FIG. 48, 49. *C. glauca* „Java'', Zygophase, 2000 × and 1800 ×.
FIG. 50. *C. glauca* „Java'', Late zygophase, 2000 ×.

to their minuteness the chromosomes in these nuclei are not suitable for further examination (see fig. 45*a*,*b*, showing a nucleus of a somatic cell and a pollen mother cell at the same magnification).

No morphological differences were to be found in the somatic chromosome-complements of *Canna glauca* „Java", „Bolivia", „Montevideo" and „Pure yellow", but it is probable that with a closer examination, after using other fixatives, they may be detected (as regards size, shape, constrictions and trabants).

MEIOTIC DIVISION. PROPHASE I: The prophase stages in these types are very similar to those described already for *Canna humilis*. The following descriptions apply to

a. Canna glauca „Java"

If penetration was good and the fixation was satisfactory, the chromomere-structure of the leptotene-threads is evident. The minute tortuous threads themselves are hardly visible; the dark

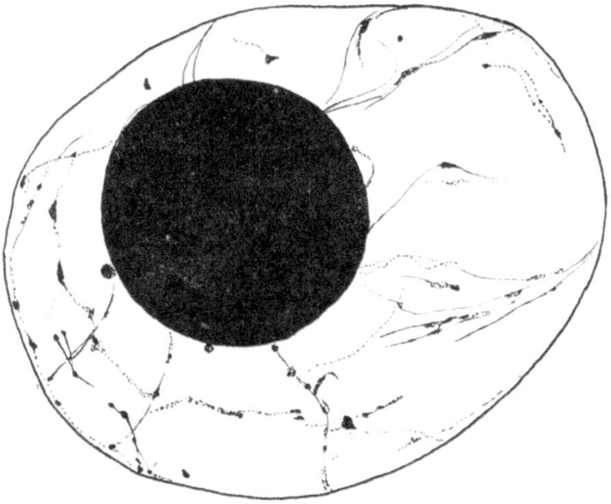

FIG. 46. *C. glauca* „Java", Leptophase, 3600 ×.

stained granules lie at unequal distances in the slightly stained threads. Both granules and threads are single (fig. 46).

These leptotene-threads associate in pairs side by side, which is to be concluded chiefly from the appearance of the so-called amphitene-stage; it is extremely difficult to determine when and where these delicate threads first come into contact with each other, but it is certainly a gradual process. In many cases the chromomeres of one thread are situated directly opposite the chromomeres on the other thread, so that this is a clear case of parasynaptic pairing (fig. 47). There is a normal zygotene and pachytene stage following this association (fig. 48–50); then the noticeably thicker thread can be seen in some places to be double, but four chromatids are not distinguishable. How the precise transition from pachytene (fig. 51) to diplotene (fig. 52) takes place is obscure. Apparently it is a rather rapid process, for suddenly the nucleus is filled up for the greater part with the diplotene threads. Sometimes their finer construction is to be examined; obviously they are tetrads, consisting of four crossed and twisted chromatids (fig. 53). At the same time the partly condensed bivalents are already visible, for the condensation does not take place simultaneously. Each of the two chromatid-pairs of a bivalent has an already condensed portion, which lie in homologous places; the other parts, still in the diplotene-stage, lie in different directions in the nuclear cavity. Only in a few cases these threads can be traced over a greater distance (fig. 53–54). By the further contraction of the bivalents (which in this species is a heterochronous phenomenon too) the real diakinesis develops; nine bivalents occur whose double nature is still quite distinct (fig. 55–56). Generally most of them are ring-bivalents; sometimes one or two rod-bivalents are developed (as was the case in *Canna humilis*).

In a few nuclei associations of two bivalents have been found (see fig. 57, 58, 59).

It cannot be decided how they have been formed; for the connections between the pairs are not quite clear; it may be in the one case (fig. 58) that two bivalents, which were totally terminalized, have come to lie parallel. If this is so, it is not a very interesting phenomenon. But such is by no means the case in the association shown by fig. 57 and 59, for here are two bivalents evidently connected by a fine, probably, double thread. It may be considered as a „secondary association'' (but this is merely a word here and not an explanation). At the end of diakinesis the nucleolus disappears; the nuclear cavity decreases. The now highly contracted bivalents lie scattered through the nucleus (fig. 60) sometimes in one half of the space and at a short

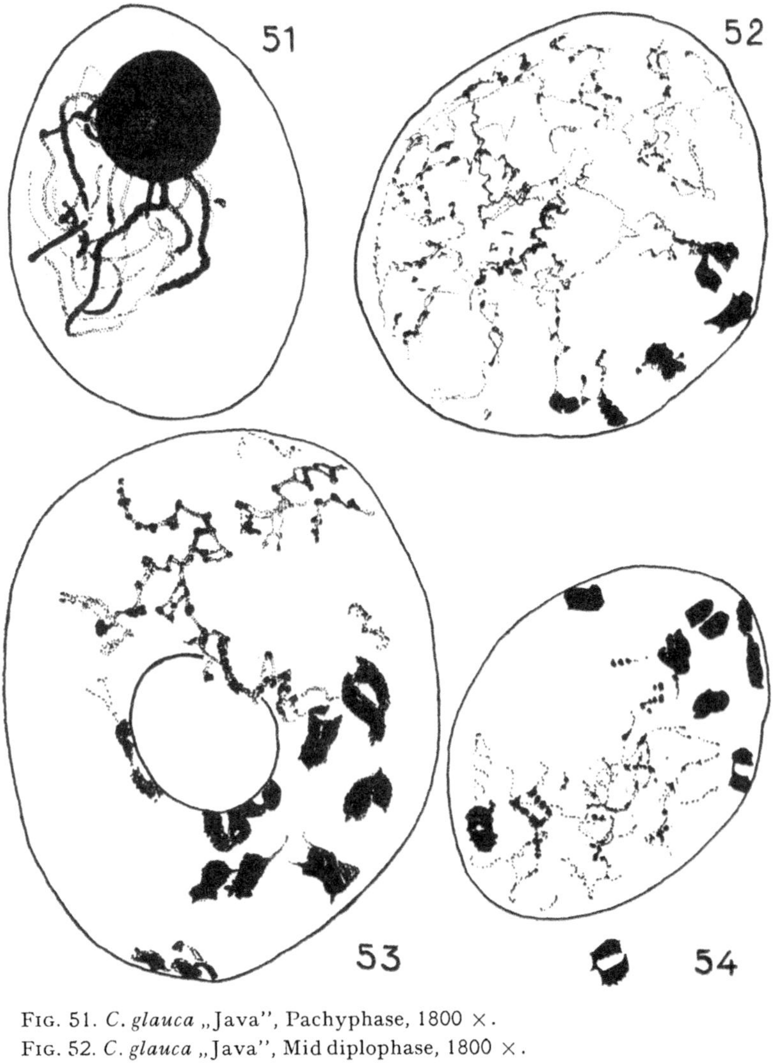

FIG. 51. *C. glauca* „Java", Pachyphase, 1800 ×.
FIG. 52. *C. glauca* „Java", Mid diplophase, 1800 ×.
FIG. 53. *C. glauca* „Java", Diplophase, 2225 ×.
FIG. 54. *C. glauca* „Java", Diplophase, 1800 ×.

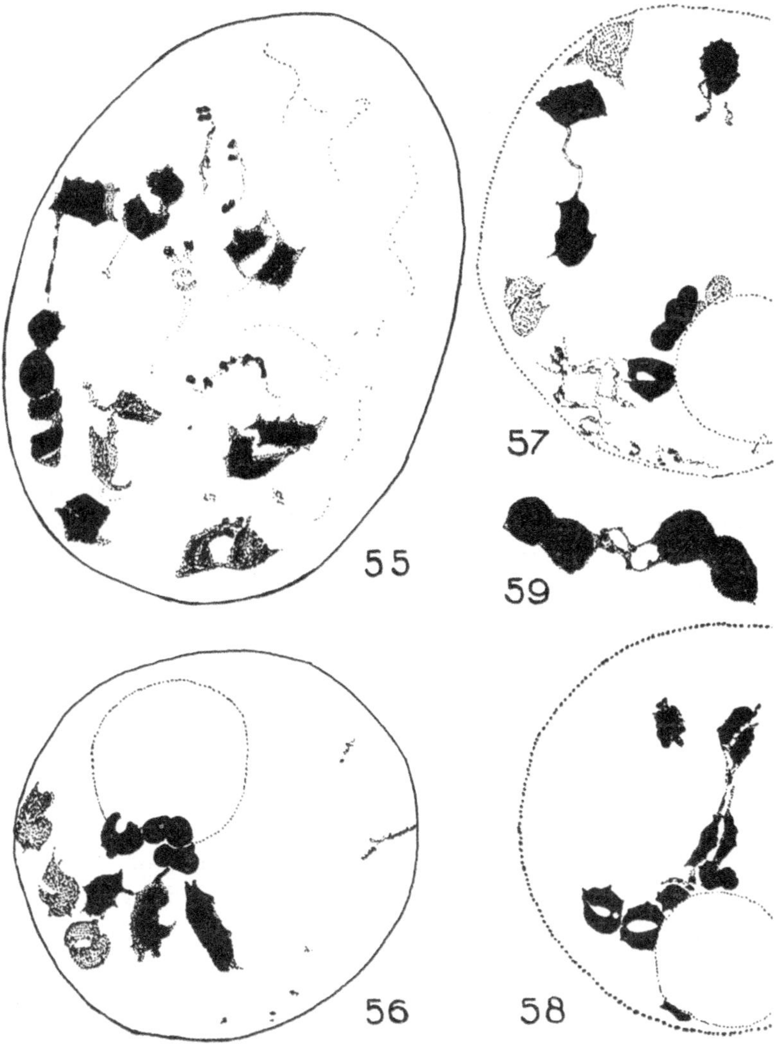

FIG. 55. *C. glauca* „Java'', Diplophase, 9 bivalents, 2700 ×.
FIG. 56. *C. glauca* „Java'', Diakinesis, 9 bivalents, 1800 ×.
FIG. 57. *C. glauca* „Java'', Early diakinesis, connected chromosomes, 2000 ×.
FIG. 58. *C. glauca* „Java'', Connected chromosomes, 1800 ×.
Fig. 59. *C. glauca* „Java'', Connected bivalents, 3600 ×.

distance from the nucleus membrane; certainly they are not of the same form and size, but it has not been possible to distinguish the individuals of the complement. In this stage and the following, when the bivalents are scattered through the spindle meanwhile developing (fig. 61–62), they may be easily counted. With *Canna glauca* „Java" the chromosomes in the bivalents are closely associated, although as a rule the double nature of the latter is to be concluded from their curved outline.

Sometimes there is a contraction following diakinesis: the chromosomes clump together in the centre of the pollen mother cell.

So far as may be judged from the observations, the spindle is a product of the nucleus; I never found in the immediate surroundings of the nucleus any indication of a multipolar spindle. But in several cases fine threads are to be seen in the nuclear-hole, which becomes more bipolar. These threads bear in some places small stained particles (with the same colour as the chromatic material). It is impossible to say, whether they are still parts of the chromosomes, or perhaps „proto-fibres" of the spindle.

METAPHASE I: The metaphase-plate is regular, but it is rather rare in the preparations, owing to the probably short duration of this stage. From side and polar views it may be inferred that the bivalents have terminal chiasmata, one or two, for they are ring- or rod-shaped. The primary constrictions are seldom quite clear.

ANAPHASE I: Although the separation of the chromosomes in early anaphase is not always simultaneous, this stage cannot be said to be irregular. Some chromosomes, possibly the rod-bivalents, may move apart earlier or more quickly (fig. 63). In this stage, counts offer no difficulty, neither in side-view nor in polar view; especially the latter give, in many cases, the impression of similarity in the grouping of the separating daughter-chromosomes. Fig. 64*a*, where the chromosomes are clearly seen passing towards the poles, gives a good notion of the anaphase I: *a*) the chromosomes separate regularly in 9+9; *b*) the double nature of them is evident (from early to mid-anaphase); *c*) not all the chromosomes of a set are in the same stage of development. In late anaphase the chromosomes are evidently split, but their appearance is somewhat different from that in early anaphase.

TELOPHASE I: When the chromosomes have reached the poles they

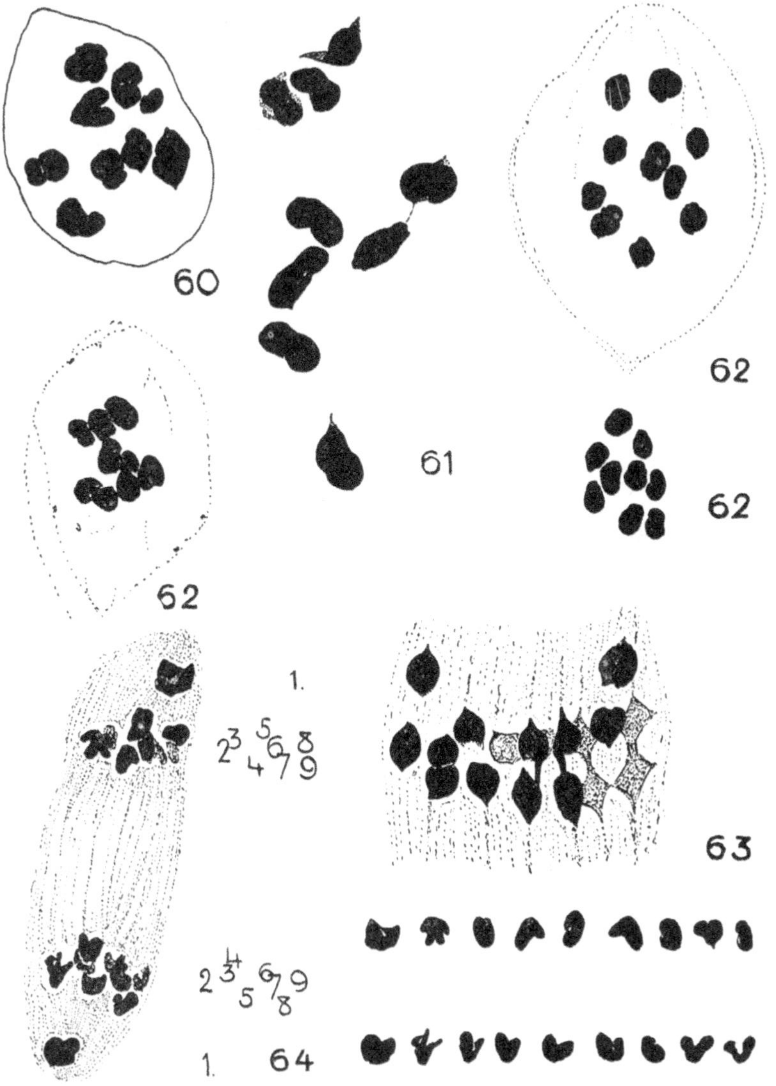

FIG. 60. *C. glauca* „Java", Diakinesis, 3000 ×.

FIG. 61. *C. glauca* „Java", Chromosomes scattered through the spindle 3600 ×.

FIG. 62. *C. glauca* „Java", *a-b-c*: chromosomes scattered through the spindle, 2700 ×.

Fig. 63. *C. glauca* „Java", Metaphase I, 3600 ×.

FIG. 64. *C. glauca* „Java", *a-b*, Anaphase I, 2000 ×.

come in contact with one another, sometimes forming a spiral-curved chain (fig. 65). Since they are rather large, counts again are easy to make, but they soon lose their sharp outlines by alveolization and the individuals are no longer distinguishable. In a transitional stage

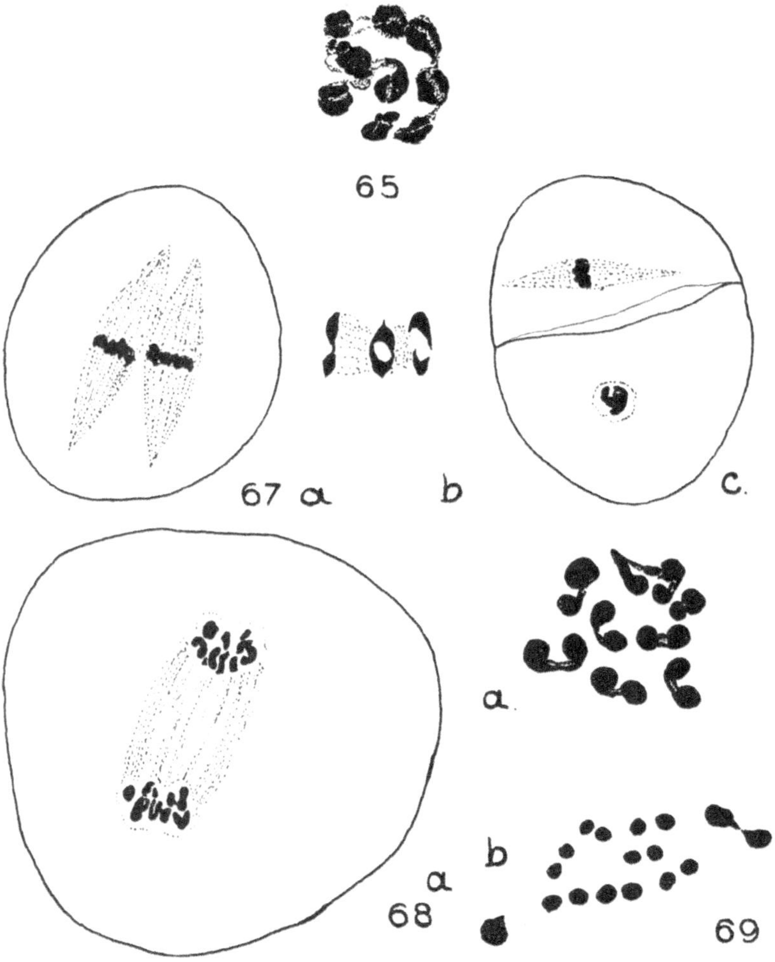

FIG. 65. *C. glauca* „Java", Late telophase, 2700 ×.
FIG. 67. *C. glauca* „Java", Second meiotic division, *a-b-c*; 1350 ×.
FIG. 68. *C. glauca* „Java", Anaphase II, regular separation of the chromosomes 900 × (compare for the number fig. 69*a-b*).
FIG. 69. *C. glauca* „Java", Anaphase II, chromosomes, top-view, 3600 ×.

from anaphase to telophase they are evidentiy double, for a split at the length can be seen, while probably the ends of the chromosomes are connected. The spindle between the daughter-nuclei extends

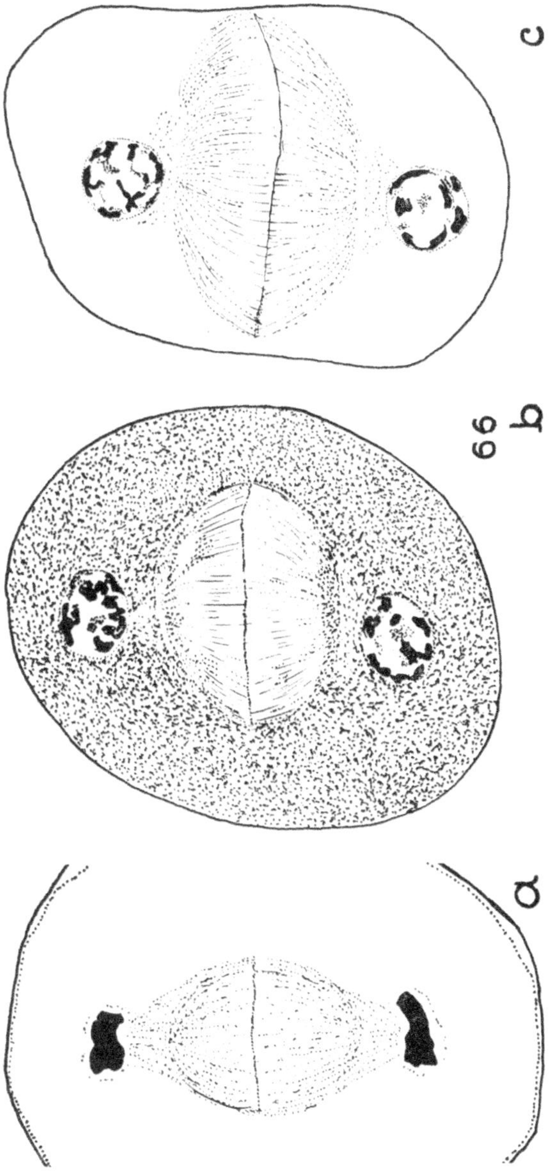

FIG. 66. *C. glauca* „Java", Telophase I, *a-b-c*; phragmoplast; 900 ×.

laterally as a phragmoplast. The first meiotic division ends with the formation of a new cell-wall, as is the rule in monocotyledonous plants (fig. 66, 67, 68b).

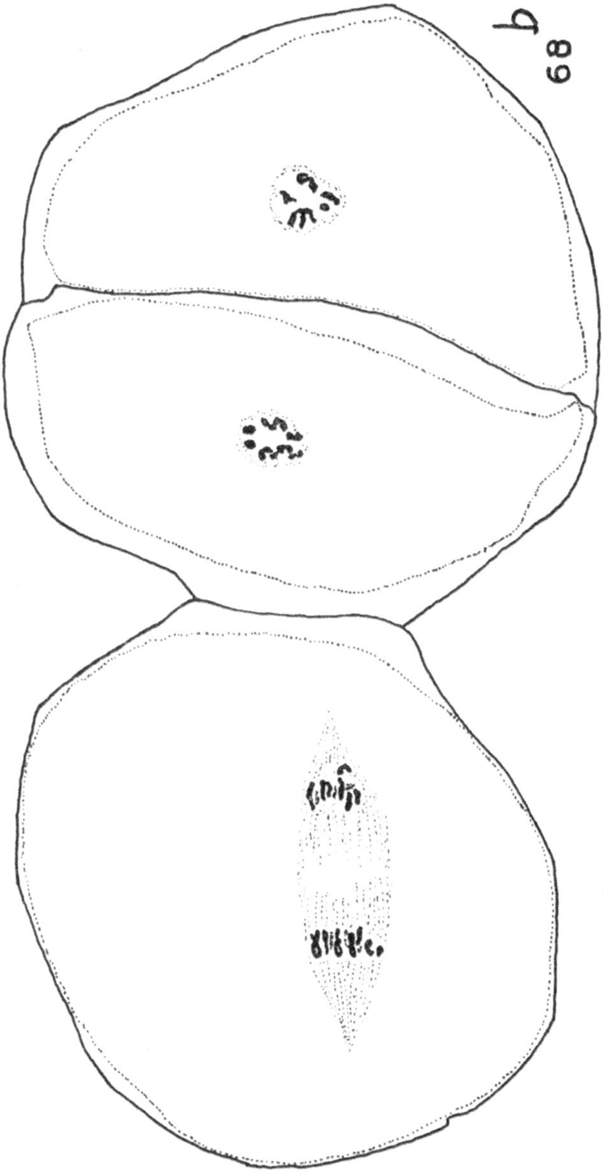

FIG. 68*b*. *C. glauca* „Java" Second meiotic division, 900 ×

SECOND MEIOTIC DIVISION: The cell-wall, separating the two daughter-cells is sometimes hardly visible; superficial examination may give the impression that no membrane at all exists.

METAPHASE AND ANAPHASE II: The spindles can lie parallel with their axes, but there are also cases where they lie cross-wise (fig. 67*a* and *c*). The causes governing such differences in position are obscure. These spindles are to be distinguished from the spindles in the first meiotic division by their shape, viz. long and narrow. A regular distribution of 9 chromosomes to each pole has been observed (fig. 69*a*). Irregular separation of chromosomes must be rare. In metaphase II the chromosomes on side view are ring-shaped (fig. 67*b*); on top-view the anaphase chromosomes have a remarkable appearance, for it gives the impression that there are too many (when namely both arms of each chromosome are clearly visible and the connection between them is vague; fig. 69).

The anaphase proceeds very regularly and changes into telophase II in which the chromosomes clump together and form a dense mass at the poles of the spindles. As a rule the new membrane develops in a regular manner and the tetrads too. If there are any irregularities in the formation of the tetrads they are rare and probably due to marring external circumstances and not to internal factors.

b. Canna glauca „Bolivia"

Concerning the development of the pollen mother cells of *Canna glauca* „Bolivia" the following facts may be mentioned: The diploid number of chromosomes = 18 (metaphase plates in somatic cells of the anthers could easily be counted, fig. 70).

After the resting-stage the nucleus passes into the leptotene-stage. The chromomeres are very clearly visible; the synaptic pairing of the leptotene-threads is evident. Zygophase-pachyphase develops as already described for *C. glauca* „Java". The diplotene-stage passes in following manner into the diakinesis: The bivalents condense gradually; the contraction begins in a certain portion of the bivalents (but in homologous parts) and this proceeds succedaneously with the still projecting diplotene threads. These spirally wounded threads contract more and more (fig. 71, 72 and 79).

Not all the bivalents of a nucleus are in the same stage of con-

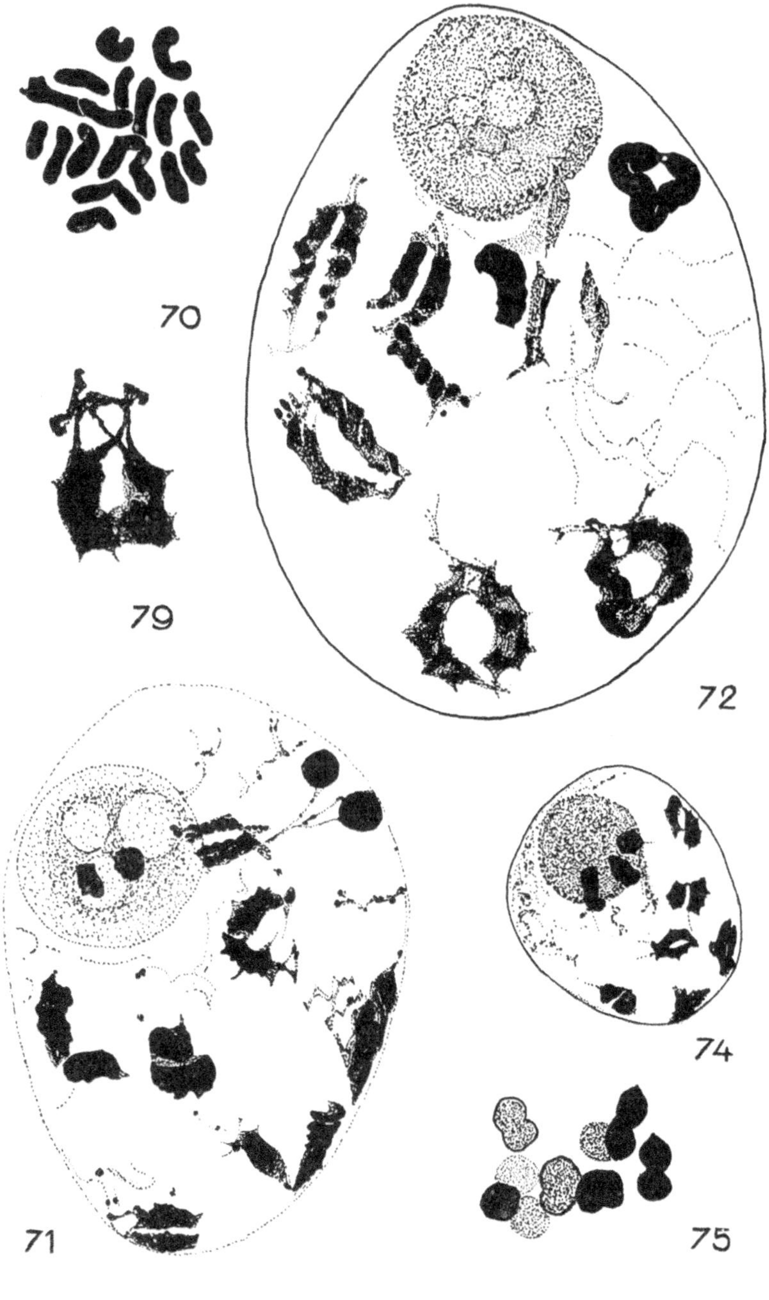

FIG. 70. *C. glauca* „Bolivia", Somatic metaphase, 3600 ×.
FIG. 71. *C. glauca* „Bolivia", Late diplophase, 2700 ×.
FIG. 72. *C. glauca* „Bolivia", Early diakinesis, 3600 ×.
Fig. 74. *C. glauca* „Bolivia", Diakinesis, 1350 ×.
FIG. 75. *C. glauca* „Bolivia", Prometaphase, 3600 ×.
FIG. 79. *C. glauca* „Bolivia", Bivalent, 3600 ×.

traction, which in this form is a heterochronous phenomenon too (fig. 72). The parts of the bivalents already condensed often lie together in another part of the nucleus than the attached non-condensed

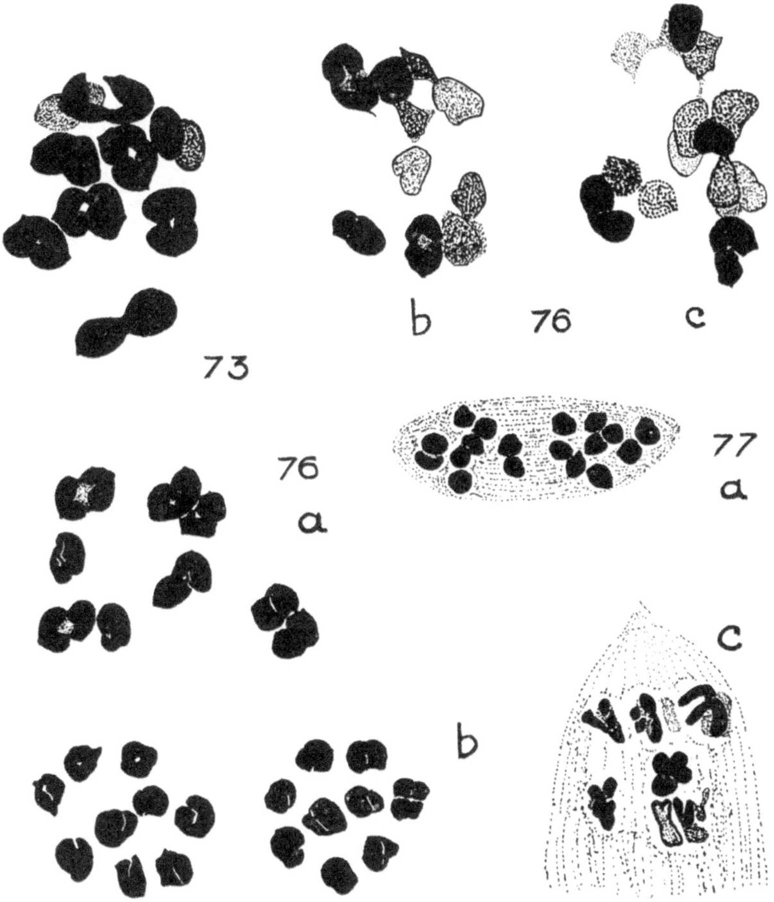

FIG. 73. *C. glauca* „Bolivia", Late diakinesis, 3600 ×.
FIG. 76. *C. glauca* „Bolivia", Chromosomes, scattered through the spindle, *a-b-c*, 3600 ×.
FIG. 77. *C. glauca* „Bolivia", Anaphase I, *a*. 1800 ×, b. 3600 ×, *c*. 3600 ×.

threads; the connection of bivalents and threads is clearly visible in „Bolivia".

At the end of diakinesis nine gemini have been developed; three of them may be found in several cases near the nucleolus, while the others lie closer to the nucleus-wall (fig. 74). The metaphase-chromosomes have as a rule two terminal connections; one of these connections may become detached first, the other being more tenacious.

When the chromosomes are scattered through the spindle, their double nature is evident although the chromosomes of each bivalent lie close together (fig. 75). The gemini may show a „secondary association" (fig. 76*a*). The spindle may be found somewhat excentrically. Obviously the transition from diakinesis to anaphase proceeds very rapidly.

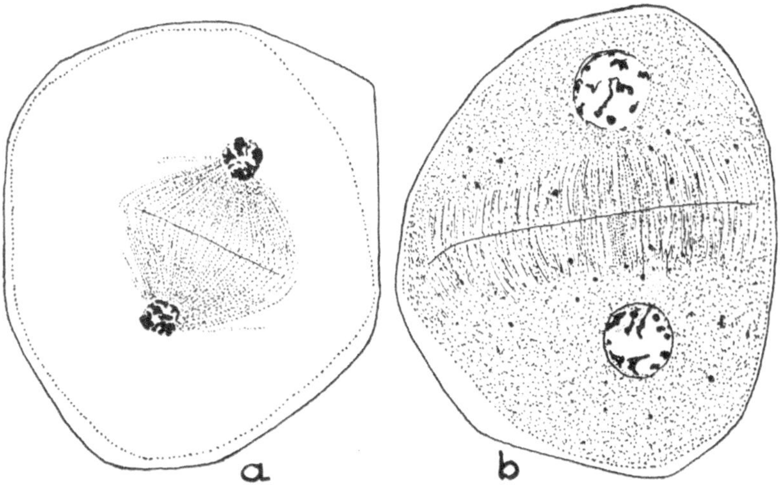

FIG. 78. *C. glauca* „Bolivia", Telophase and dyad, 900 ×.

The separation of the chromosomes in anaphase is regular: 9 chromosomes pass to each pole, but this does not take place simultaneously (fig. 77). In mid and late anaphase the chromosomes are clearly double; in late anaphase they are somewhat spiral-formed. For the rest see: *C. glauca* „Java" and the illustrations, belonging to *C. glauca* „Bolivia" (fig. 78*a* and *b*).

c. *Canna glauca* „Montevideo"

This variety requires no detailed description for the same results were obtained from the investigations as with the forms already described. The pairing is evidently parasynaptic (fig. 80).

FIG. 80. *C. glauca* „Montevideo", Pachyphase, 1800 ×.

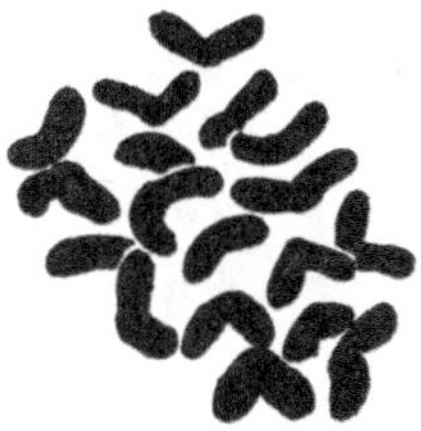

FIG. 81. *C. glauca* „Pure yellow"; Somatic metaphase, 3600 ×.

d. Canna glauca „Pure yellow"

The diploid number is 18. (Somatic cells of the anthers, fig. 81). Clear differences with the other forms examined, as regards the meiotic phenomena, are not to be seen.

E. *Canna glauca* „Java" × *discolor* F_1-hybrid

Undoubtedly, in the somatic tissue of the anthers the metaphase plate clearly shows 18 chromosomes (fig. 82). It is very probable that morphological differences occur, but owing to the unsuitable fixing fluid they were not so clearly developed that details could be analyzed. The preparations available did not show in the early prophase the fine chromomere-structure of the leptotene-threads, but we may venture to suppose that the zygotene-pairing takes place in the same manner as it has already been described for *Canna humilis* and the other forms. (For the appearance of zygotene-pachytene is essentially the same as that of the other forms described). Apparently these stages pass gradually from one to the other, but not simultaneously for all the chromosomes of a pollen mother cell.

The ultimate zygotene threads, lying in large loops in the nuclear cavity are undoubtedly double: often they lump together in the proximity of the nucleolus. The pachytene and the diplotene develop in the manner mentioned. The individual chromosomes are not to be traced, so that a detailed study of this stage, namely accurate count-

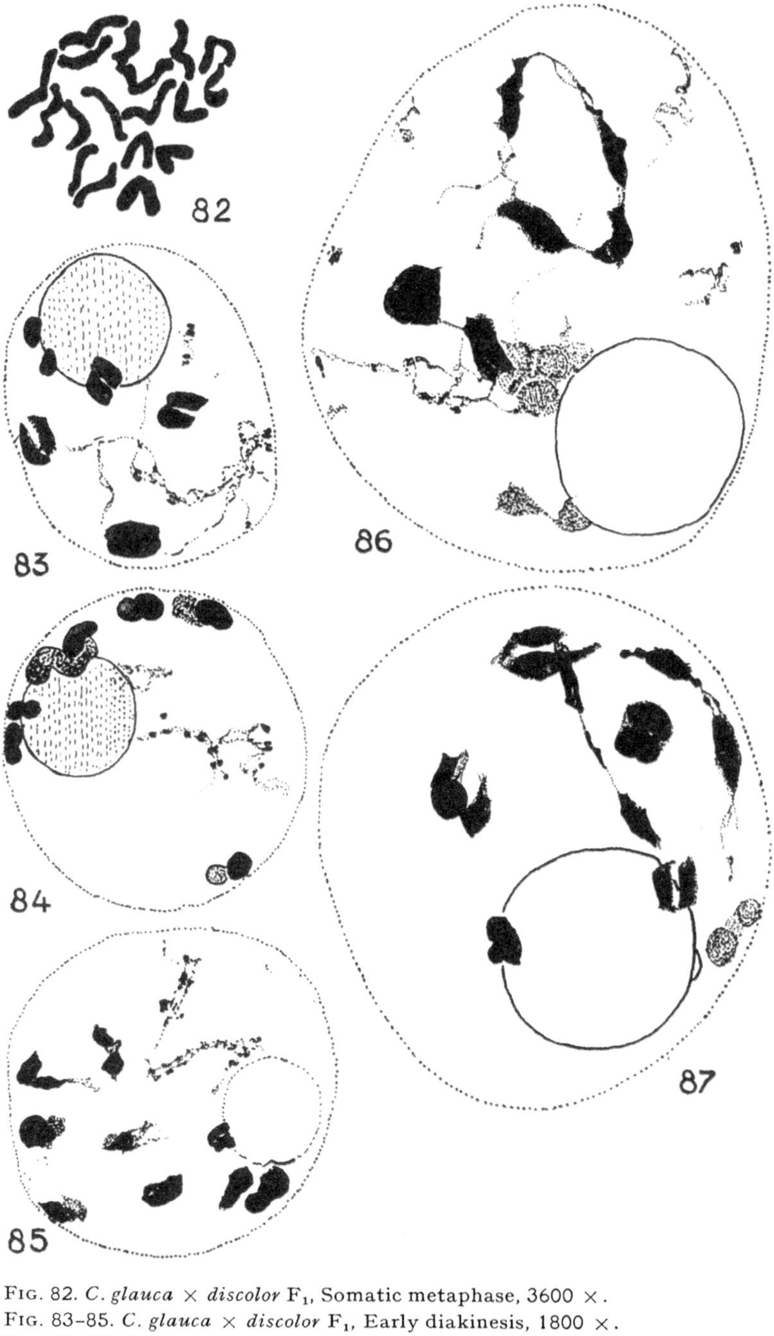

FIG. 82. *C. glauca* × *discolor* F_1, Somatic metaphase, 3600 ×.
FIG. 83–85. *C. glauca* × *discolor* F_1, Early diakinesis, 1800 ×.
FIG. 86–87. *C. glauca* × *discolor* F_1, Bivalents and chains in the same nucleus.

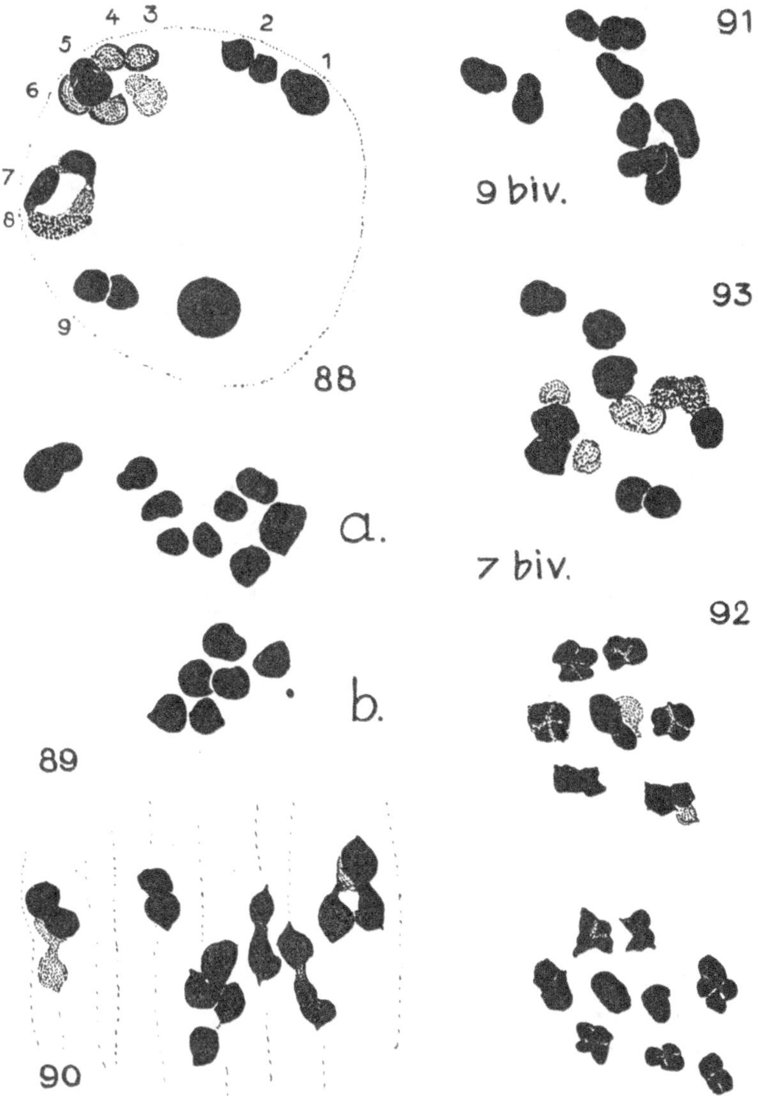

FIG. 88. *C. glauca* $\times$ *discolor* F_1, Diakinesis, 9 sivalents, 3600 $\times$
FIG. 89. *C. glauca* $\times$ *discolor* F_1, Anaphase I, 3600 $\times$.
FIG. 90. *C. glauca* $\times$ *discolor* F_1, Metaphase I, 3600 $\times$.
FIG. 91. *C. glauca* $\times$ *discolor* F_1, Nine bivalents, 3600 $\times$.
FIG. 92. *C. glauca* $\times$ *discolor* F_1, Anaphase I, regular separation.
FIG. 93. *C. glauca* $\times$ *discolor* F_1, Seven bivalents and four univalents, 3600 $\times$

ing of chiasmata is impossible (as in so many other cases). The chromosomes become distinguishable at the beginning of diakinesis, while in mid diplotene the condensation of some bivalents is already

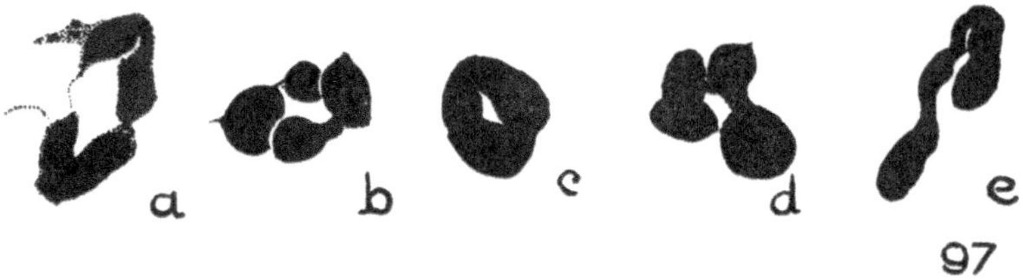

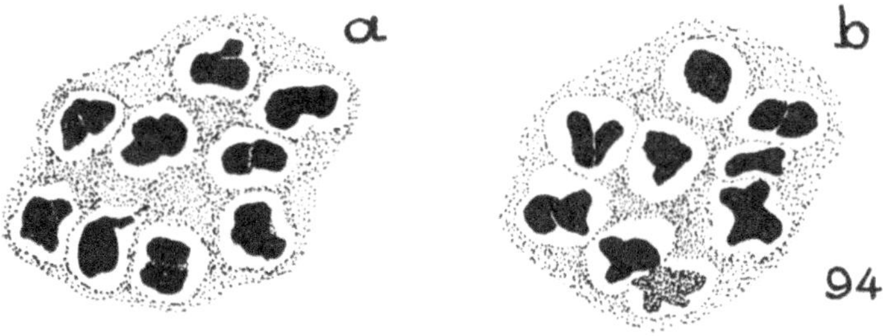

FIG. 94. *C. glauca* × *discolor* F_1, Anaphase I, similar grouping of the chromosomes, ± 3600 ×.
FIG. 95. *C. glauca* × *discolor* F_1, Late anaphase I, 2400 ×.
FIG. 96. *C. glauca* × *discolor* F_1, Mid anaphase I, 2400 ×.
FIG. 97. *C. glauca* × *discolor* F_1, Multivalent connections.

clearly visible (fig. 83); in late diplotene the greater part of the threads again are seen to be double and paired in most of the nuclei

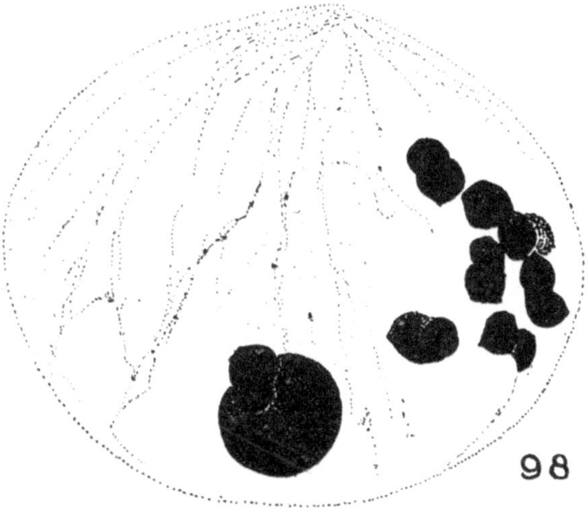

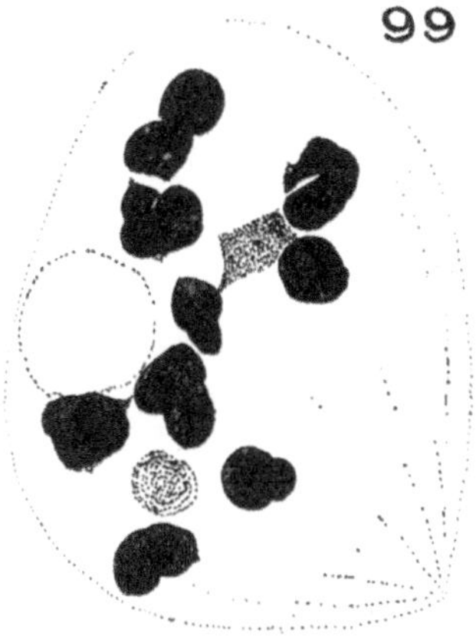

FIG. 98–99. *C. glauca* × *discolor* F_1, Spindle-development in late diakinesis.
FIG. 101. *C. glauca* × *discolor* F_1, Abnormal spindle.

(fig. 84, 85). But generally the exact analysis of the nuclei is difficult, owing to the fact that the gemini as a rule are found in the same part of the nucleus, so that they often cover each other, wholly or partially. In the course of diakinesis the association of the chromosomes gives the impression that nine pairs will be formed (fig. 91); it may be that some are incompletely paired for they approach each other without uniting (fig. 93, 88). Univalents lying wide apart are to be seen in a few cases only. In an advanced stage of diakinesis pollen mother cells sometimes still contain a great number of probably unpaired threads; the chromomeres are larger than those of the leptotene threads. Possibly the former threads may develop into univalents. In metaphase I pairing is almost complete, although in some bivalents the chromosomes may be only loosely connected.

If one has a clear notion of the structure of the chromosomes in late anaphase and early telophase, they may easily be counted. Their structure is equal to that of the chromosomes of the different Canna types described in the same stages (fig. 96, 95).

The further development follows the usual course and needs no more elucidation (fig. 100; metaphase and anaphase II).

VII. DISCUSSION

Terminology. Present-day terminology for the prophase-stages is not entirely suitable for the facts described here. Properly speaking, the terms leptotene, zygotene, etc. cannot be used (in many cases) in their original significance, seeing that these stages are often not distinctly separated, but gradually change from one into the other. The difficulty of denominating the transitional stages has already been felt by many writers which has given rise to expressions such as amphitene, mid-zygotene, pro-metaphase, etc.

With *Canna*, there is no definite zygotene stage for the whole nucleus; often paired and single threads are to be seen in the same nucleus, a stage, equivalent to amphitene of JANSSENS and to mid-zygotene of other researchers. The words of HUSKINS and SMITH (1935): „And pachytene likewise is not a definite stage for the whole nucleus'' are applicable for *Canna* too. The same holds good for diplotene and diakinesis.

It would be recommendable to choose new terms, more expressive of the dynamic character of the phenomena, but present-day terminology has become too popular in cytology to admit of other terms.

Fixation. Although some researchers object to the use of Carnoy's fluid (DARLINGTON, 1932, p. 491) it was successfully used in this investigation and late literature creates the impression that a great many cytologists again resort to the use of this fixative.

On comparing the results of investigations notice should be taken of the fixation-fluids used for the treatment, as it has appeared from this investigation, too, that the results may be very different, owing to the fixatives used (p. 18). If the above is not taken into account it may lead to misconceptions, especially of the finer structures. It is possible that in some cases wrong conclusions were drawn, e.g. with regard to prochromosomes and telosyndetic pairing, owing to unsuitable fixatives. It is worth mentioning the words of GATES and GOODWIN (1931): ,,The papers of LATTER (1926) and of MAEDA (1930) on *Lathyrus*, also lead to the conclusion that according to the method of fixation one may obtain an essentially telosynaptic or parasynaptic story from preparations, the fusion of delicate parallel threads in certain fixatives giving an appearance of telosynapsis".

Prochromosomes. In the pollen mother cells of the species examined no prochromosomes could be detected, (neither in the embryo-sac mother cells observed) although, on the contrary, KRACAUER (1930) does state the existence in the embryo-sac mother cells of *Canna indica*, but he does not say so of the pollen mother cells.

Method of pairing. Definite evidence of parasynapsis in *Canna* has been found. The pairing of the homologous chromosomes in Canna does not take place simultaneously along their whole length, a phenomenon which is also observed with other plants: some threads being partly paired and partly unpaired, the latter conspicuously thinner than the paired ones.

It is open to doubt, whether contact originates accidentally in any place or whether it is a regular process. Counting or estimating the number of chromomeres in a nucleus, which has appeared to be possible for species of *Lilium* and *Trillium* (BELLING, 1928, 1931; HUSKINS and SMITH, 1935), is impossible with *Canna*.

That a continuous spireme develops is not excluded, yet highly improbable. When the chromosomes in early prophase are long, as is the case with *Canna*, free ends are difficult to observe. In favourable cases the loops in pachytene gave the impression of consisting of four strings; it is, therefore very probable that in this stage four distinct chromatids are developed.

Condensation. In what manner the pachytene changes into the diplotene-stage is not quite clear. In particular, it has not been possible to discover with certainty the development of the connections between chromatids and chromatid-pairs, i.e. the formation of the chiasmata. Indicating the places and the number of the chiasmata and the change in their position, is very difficult or hardly possible, as the bivalents do not contract simultaneously along their whole length. Locally this process is more in advance; probably this more condensed part does not lie in the middle, but (at any rate for some of the bivalents) more towards one of the ends, the distal parts are then still in the diplotene stage. There is a relation between the quantity of diplotene threads and the contracted parts of the bivalents. This very gradual decrease of the mass of diplotene threads and the gradual increase of the condensing portions of the bivalents is probably not so conspicuous in other species.

(Surely this phenomenon cannot be the consequence of bad fixation, as it appears after different treatments and with different species and varieties, at different times collected).

Only occasionally similar observations are met with in literature, but these phenomena are however not quite the same as those described for *Canna*.

CATCHESIDE (1931) describes a comparable phenomenon for *Oenothera*: „the chromosomes have the major portion of their substance located between two chiasmata, suggesting that (1) chiasma formation is most frequent towards the ends of the bivalents, and that (2) there is a movement of chiasmata away from this portion towards the much more slender distal portions". „They are very variable in thickness and outline, being thick at their centres and thinner distally, the surface in the thick portion is more or less irregular, while in the thinnest distal portions an appearance suggestive of a row of variously sized chromomeres may be distinguished in some of the less condensed bivalents".

LATTER (1932) in a description of the meiotic divisions in the pollen mother cells of *Malva sylvestris* mentions, that an extreme contraction of the bivalents occurs immediately prior to diakinesis, and that the appearance of the nuclei in this stage suggests that a portion of the thread mass is not included in the chromosomes. In *Malva sylvestris* these remaining threads form a faintly staining mass of somewhat reticulate structure around the nucleolus, while the contracted bivalents take up a peripheral position. In all diakinetic nuclei a quantity of this faintly-stained substance as well as the bivalent chromosomes has been observed by LATTER. But comparatively few stages of diakinesis were found in the material examined. LATTER supposes that this remaining portion probably corresponds to the linin-threads, described in *Lavatera* by BYXBEE, which, according to her account, contribute to spindle formation. In *Malva sylvestris* these faintly-stained indistinct strands of substance are present in the nucleus in addition to the bivalent chromosomes still in late diakinesis; according to LATTER they were not utilized at chromosome-formation. With *Malva sylvestris* there is evidently no clear connection between the threads and the bivalents in diakinesis, the structure of the threads cannot be inferred with certainty from the illustrations. With the species of the genus *Canna* examined, the threads mentioned are gradually taken up into the chromosomes and therefore the phenomenon is comparable with that in *Oenothera*, as described by CATCHESIDE.

HUGH DAVIE (1934) studying *Lavatera*, found also, that in diakinesis the appearance of the nuclei suggests that not all the chromatic material has been used up in the formation of the bivalent chromosomes.

Spindle. BYXBEE supposes, the threads contribute to spindle formation; LATTER had insufficient material for an accurate study, but yet he has a different notion of the mode of spindle formation. BYXBEE describes the cytoplasm of *Lavatera* as composed of two constituents, a fibrous network and a granular substance about the nuclear wall, which leads to the formation of the characteristic „perinuclear zone". The fibres grow into the nuclear cavity and mingle with the „linin-threads". Surely this is not the case with *Canna*; the „threads" cannot be brought into relation with the formation of the spindle, although in some cases it was clearly observable in late diakinesis that the nuclear-cavity had become spindle-shaped and within threads were visible, bearing small, stained particles (fig. 21, 98, 99). A multipolar spindle or the inward growth of spindle fibres has never been observed with the *Canna*-species examined. These observations support the view that the spindle arises from the nucleus, but there is no definite evidence, that it originates in a portion of the threads with chromatic material.

Although rare, abnormal spindles have been found (fig. 34, fig. 101); in all cases the chromosomes lie within the karyoplasm of the spindle.

Sometimes, each chromosome in metaphase and anaphase seems to lie in a separate part of the spindle (fig. 94*a*, *b*). Some investigators consider this to be an artefact, but to others it is an indication that the spindle is a compound structure consisting of elements, related to the chromosomes. Although such cells and spindles look very normal, this, however, does not mean that in living cells the same structure would be observable. It may be that the spindle-fibres must be considered as artefacts (cf. BĚLAŘ 1928; BLEIER 1930; H. DAVIE 1934) and likewise it may be for the described anaphase-pictures. (The illustration fig. 94*a* and *b* shows clearly the similarity of the grouping of the chromosomes).

Associations. The metaphase of the first meiotic division suggests that some of the chromosome-bivalents show secondary associations; some figures in a paper of HONING, too, (Canna Crosses II), give this impression. If there is actual secondary association, the whole chromosome-complement is composed of 3 bivalents and 3×2 (more related) bivalents and it must be possible, that some species are polyploids. But there are no other indications in support of this suggestion; species with a smaller number of chromosomes than 18 have not been found so far. Neither have particularities been published about the pairing in triploid forms, which might elucidate this phenomenon.

As a rule meiosis proceeds very regularly with clear formation of nine bivalents, this is worth mentioning, as most of the cultivated Cannas (and presumably also many so-called „wild'' forms) are species-hybrids.

Coupling of factors, found by HONING, cannot be ascribed to chromosome-coupling, connections as rings or chains are of rare occurrence and it is quite well possible that only pseudo-chains are formed (by secondary association).

It is probable that a comparative study of the development of the diploid, triploid and gigas-forms may shed more light on the subject.

VIII. SUMMARY

A. 1. With *Canna humilis* the chromosomes associate in pairs: it is a clear case of parasyndesis.

2. Though the chromomeres are clearly visible (after suitable fixation) their number cannot be definitely fixed.

3. It is probable that in the diplotene-stage interstitial chiasmata are formed, which become completely terminalized.

4. The condensation of the bivalents during diplotene-diakinesis proceeds gradually and heterochronously.

5. In diakinesis and metaphase I multivalents are totally absent.
6. The diploid number is 18; somatic pairing is not prominent.
7. The haploid number is 9, which is clearly evident from diakinesis, metaphase and anaphase.

In anaphase I and also in anaphase II the chromosomes separate regularly in 9+9.

8. Dyads and tetrads are formed in a regular manner.
9. A certain portion of the pollen mother cells perishes already before division; the causes are not to be ascertained.
10. Of the abundance of pollen formed, only a very small percentage is sterile.

B. The other forms require no further description here for almost the same results were obtained from the investigations.

My best thanks are due to Prof. Dr. J. A. HONING for his interest in my work and for valuable criticism in the preparation of this paper.

IX. BIBLIOGRAPHY

BAKER, JOHN R., 1933. Cytological technique. London.

BAUR, E., 1911. Einführung in die Vererbungslehre. Berlin.

BĚLAŘ, K., 1928. Die Cytologischen Grundlagen der Vererbung, Berlin.

BELLING, J., 1921. The behaviour of homologous chromosomes in a triploid Canna. Proc. Nat. Ac. Ss. Washington, Bd. 7.

BELLING, J., 1921. On counting chromosomes in pollen mother cells. Amer. Natur. 55.

BELLING, J., 1925. *a*. Production of triploid and tetraploid plants. J. Hered. Vol. 16.

BELLING, J., 1925. *b*. Chromosomes of Canna and of Hemerocallis. J. Hered. Vol. 16.

BELLING, J., 1926. The iron-acetocarmine method of fixing and staining chromosomes. Biol. Bull. 50.

BELLING, J., 1928. A method for the study of chromosomes in pollen mother cells. Univ. Calif. Pub. Bot. 14.

BELLING, J., 1930. The use of the microscope. New York—London.

BELLING, J., 1931. Chromomeres of Liliaceous plants. Univ. Calif. Pub. Bot. 16.

BLEIER, H., 1930. Untersuchungen über das Verhalten der verschiedenen Kernkomponenten bei der Reduktionsteilung von Bastarden. La Cellule 40.

BLEIER, H., 1931. Zur Kausalanalyse der Kernteilung. Genetica 13.

BYXBEE, E. S., 1900. The development of the karyokinetic spindle in the pollen mother cells of Lavatera. Proc. Calif. Acad. Sci. Ser. 3 (Bot.) 2.

CATCHESIDE, D. G., 1931. Critical evidence of parasynapsis in Oenothera. Proc. Roy. Soc., B. 109.

COSTERUS, J. C., 1916*a*. Canna's. De Natuur, 36e jaarg.

COSTERUS, J. C., 1916*b*. A fresh investigation into the structure of the flower of Canna. Ann. Jard. Bot. Buitenzorg. 2 Serie, Vol. XIV.

COSTERUS, J. C., 1916*c*. Die Übereinstimmung und der Unterschied in dem Bau der Blumen von Canna und derjenigen der Marantaceen. Ann. Jard. Botan. Buitenzorg. 2 Serie, Vol. XV.

COSTERUS, J. C., 1920. La structure de la fleur du Canna. Rec. trav. bot. Neerl. XVII.

DARLINGTON, C. D., 1932. Recent advances in Cytology. London.

DAVIE, J. Hugh, 1934. Cytological studies in the Malvacae and certain related families. Journ. of Genetics. Vol. XXVIII.

GAISER, L. O., 1926. A list of chromosome numbers in Angiosperms. Genetica 8.

GAISER, L. O., 1930. *a*. Chromosome numbers in Angiosperms II. Bibliogr. Genet. 6.

GAISER, L. O., 1930. *b*. Chromosome numbers in Angiosperms III. Genetica 12.

GATES, R. R. and GOODWIN, K. M., 1931. Meiosis in Oenothera purpurata and Oe. blandina. Proc. Roy. Soc., B. 109.

GEITLER, L., 1935. Grundriss der Cytologie. Berlin.

GRÉGOIRE, V., 1912. Les phénomènes de la métaphase et de l'anaphase, etc. Ann. Soc. Sci. Bruxelles. 36.

HEITZ, E., 1926. Der Nachweis der Chromosomen. Zeitschr. f. Bot. 18.

HEITZ, E., 1929. Heterochromatin, Chromocentren, Chromomeren. Ber.d.d. bot. Ges. 47.

HONING, J. A., 1915. Kreuzungsversuche mit Canna-varietäten. Rec. trav. bot. Neerl. 12.

HONING, J. A., 1923. Canna Crosses I. Meded. Landb. Hoogeschool, Wageningen. Vol. 26. Paper 2.

HONING, J. A., 1928. Canna Crosses II. Meded. Landb. Hoogeschool, Wageningen. Vol. 32. Paper I.

HUSKINS, C. LEON., and STANLY G. SMITH, 1935. Meiotic chromosome structure in Trillium erectum L. Ann. of Bot., Vol. XLIX.

KOERNICKE, M., 1903. Der heutige Stand der pflanzlichen Zellforschung. Ber d. d. bot. Ges. Bd. 21.

KRACAUER, PAUL, 1930. Die Haploidgeneration von Canna indica L. Berlin.

KRÄNZLIN, F., 1912. Engler, Pflanzenreich IV, 47. (Cannaceae) Leipzig.

LA COUR, L. F., 1931. Improvements in everyday technique in plant cytology. Journ. Roy. Micr. Soc. 51.

LAWRENCE, W. J. C., 1931. The secondary association of chromosomes. Cytologia. 2.

LELIVELD, J. A., 1931. Some remarks on the cytology of Oenothera. Rec. trav. bot. néerl. Vol. 25.
LINSBAUER, K., 1921—1934. Handb. der Pflanzenanatomie. Band I, Band II, Band X 2.
DE MEYERE, J. C. H., 1928. Über haltbare, rasche Färbung vermittels Azeto-Karmin. Zeits. f. Wiss. Mikr. u. f. mikr. Technik Bd. 46.
DE MEYERE, J. C. H., 1930. Über einige europäische Insekten, besonders günstig zum studium der Reifungsteilungen, nebst einigen Zusätzen zur Azeto-karmin Methode. Zool. Anz. Bd. 88.
Mc.CLUNG, C. E., 1929. Handbook of microscopical technique. New York.
SCHNARF, K., 1931. Vergleichende Embryologie der Angiospermen. Berlin.
SCHÜRHOFF, P. N., 1926. Die Zytologie der Blütenpflanzen. Stuttgart.
TISCHLER, G., 1920/22, 1934. Allgemeine Pflanzenkaryologie. Berlin.
TISCHLER, G., 1927. Pflanzliche Chromosomen-Zahlen. Tabulae biologicae Bd. IV,
TISCHLER, G., 1931. Tabulae biol. Period. I.
TOKUGAWA I., and KUWADA Y., 1924. Cytol. stud. on some garden-varieties of Canna. Jap. Journ. of Bot. Bd. 2.
WETTSTEIN, R., 1935. Handb. der syst. Botanik. Leipzig.
WIEGAND, K. M., 1900. The development of the embryosac in some monocotyledonous plants. Bot. Gazette. Bd. 30.
WINKLER, H., 1930. Engler-Prantl. Natürl. Pflanzenfam. Bd. 15*a*.

STELLINGEN

I

De spoel ontstaat uit de kern.

II

De term synapsis=synezisis is overbodig.

III

De terminologie voor de beschrijving van de reductie-deeling is niet meer in overeenstemming met de bekende feiten.

IV

De beschouwingen van SANDS bewijzen niet, dat bij *Oenothera* telosyndese voorkomt (Amer. Journ. of Bot., Vol. 21, 1934).

V

Ten onrechte beschouwt TSCHERMAK als bewezen, dat de matrokliene nakomelingen in kruisingen van de geslachten *Pisum, Lens, Vicia* en *Ervillea* berusten op "hybridogene parthenogenesis" (Der Züchter, Jahrg. 7, 1935).

VI

De chromosomen-morphologie heeft, voorloopig althans, zeer geringe waarde voor de systematiek.

VII

De zg. glaciaal-relicten onder de hoogere planten van ons land worden ten onrechte als zoodanig beschouwd.

VIII

De proeven van VAN DER PAAUW, met betrekking tot de opname van water onder omstandigheden die de transpiratie verhinderen, bevestigen de juistheid van de theorie van SEN en BLACKMAN (Recueil d. trav. bot. néerl., Vol. 32, 1935).

IX

Het onderzoek van BURGEFF maakt het waarschijnlijk, dat de gunstige invloed van suikers op de kieming van Orchideeën-zaden in reincultuur gedeeltelijk berust op de aanwezigheid van oligodynamisch werkzame stoffen (Ber. dts. bot. Ges., Bd. 52, 1934).

X

Er is geen reden "de lijn van WALLACE" geheel te verwerpen.

XI

De grondige behandeling van de planten-morphologie is voor de leerlingen van de lagere klassen van inrichtingen voor middelbaar en gymnasiaal onderwijs van groote waarde.

XII

Een gepaste afwisseling van klassikaal onderwijs en individueele zelfwerkzaamheid is voor het onderwijs in de biologie het meest gewenscht.

XIII

Het "schoonheids-element" in de levende natuur behoort bij het onderwijs niet te worden vergeten.

XIV

Het zg. hospiteeren is voor den a. s. leeraar de belangrijkste voorbereiding voor het zelfstandig lesgeven.

GPSR Compliance

The European Union's (EU) General Product Safety Regulation (GPSR) is a set of rules that requires consumer products to be safe and our obligations to ensure this.

If you have any concerns about our products, you can contact us on ProductSafety@springernature.com

In case Publisher is established outside the EU, the EU authorized representative is:

Springer Nature Customer Service Center GmbH
Europaplatz 3
69115 Heidelberg, Germany

Batch number: 10824490

Printed by Printforce, the Netherlands